AF553106

Descriptive and Objective Insect Taxonomy

NIPA GENX ELECTRONIC RESOURCES & SOLUTIONS P. LTD.
New Delhi-110 034

www.nipabooks.com

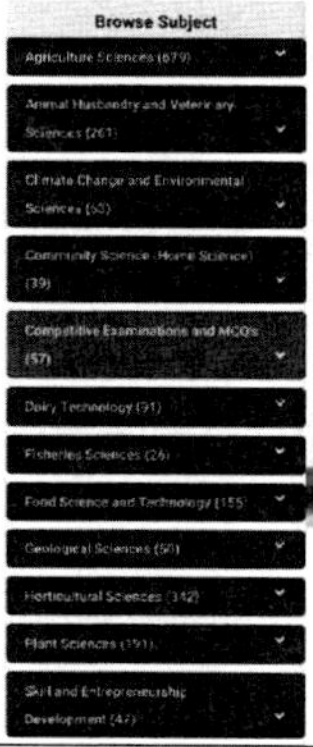

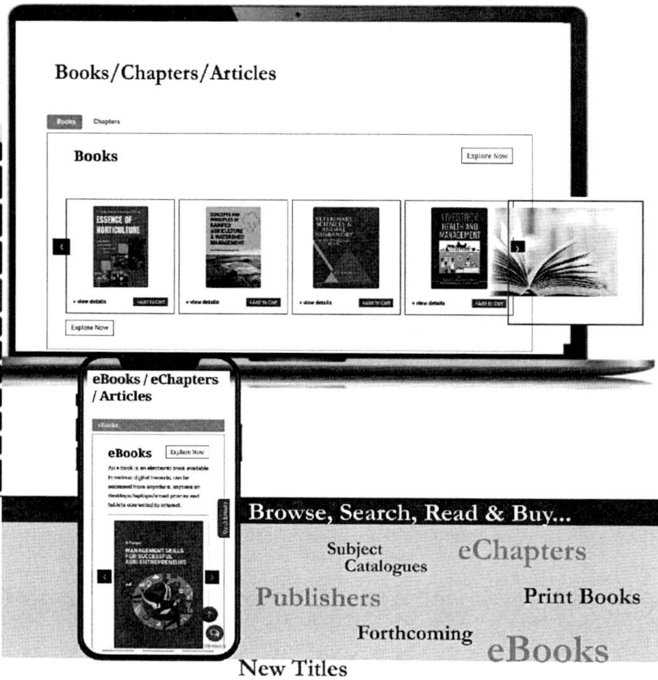

nipa

Books/Chapters/Articles

Books

Explore Now

eBooks / eChapters / Articles

eBooks

Browse, Search, Read & Buy...

Subject Catalogues

eChapters

Publishers

Print Books

Forthcoming

eBooks

New Titles

scan for catalogue

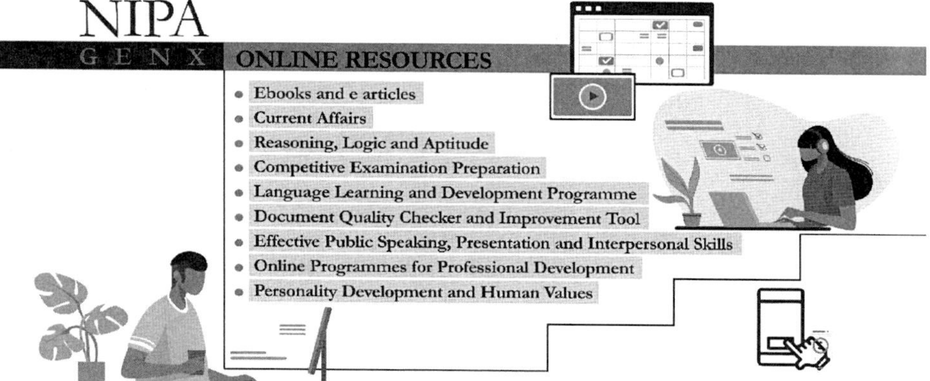

Descriptive and Objective Insect Taxonomy

Anil Kumar
Assistant Professor-cum-Scientist
Department of Entomology
Sugarcane Research Institute
Dr. Rajendra Prasad Central Agricultural University
PUSA-848 125, Samastipur, Bihar, India

NIPA GENX ELECTRONIC RESOURCES & SOLUTIONS P. LTD.
New Delhi-110 034

NIPA GENX ELECTRONIC RESOURCES & SOLUTIONS P. LTD.

101,103, Vikas Surya Plaza, CU Block
L.S.C.Market, Pitam Pura, New Delhi-110 034
Ph : +91 11 27341616, 27341717, 27341718
E-mail:newindiapublishingagency@gmail.com
www: www.nipabooks.com

For customer assistance, please contact
Phone: + 91-11-27 34 17 17 Fax: + 91-11- 27 34 16 16
E-Mail: feedbacks@nipabooks.com

© 2022, Author

ISBN: 978-93-91383-52-7

All rights reserved. No part of this publication may be reproduced, stored in a retrieval system or transmitted in any form or by any means, including electronic, mechanical, photocopying recording or otherwise without the prior written permission of the publisher or the copyright holder.

This book contains information obtained from authentic and highly reliable sources. Reasonable efforts have been made to publish reliable data and information, but the author/s, editor/s and publisher cannot assume responsibility for the validity, accuracy or completeness of all materials or information published herein or the consequences of their use. The work is published with the understanding that the publisher and author/s are not attempting to render any professional services. The author/s, editor/s and publisher have attempted to trace and acknowledge the copyright holders of all material reproduced in this publication and apologize to copyright holders if permission and/or acknowledgements to publish in this form have not been taken. If any copyrighted material has not been acknowledged, please write to us and let us know so that we may rectify the error, in subsequent reprints.

Trademark Notice: NIPA, the NIPA logos and their presentations (the way they are written/ presented) in this book are the trademarks of the publisher and hence may not be used without written permission, if copied or used without authorization, the infringer will be prosecuted as per law.

NIPA also publishes books in a variety of electronic formats. Some content that appears in print may not be available in electronic books, and vice versa.

Composed and Designed by NIPA.

Preface

Systematics is the backbone of all life science studies as it provides essentials to study biology. The subject matter forms the tool to study biodiversity. Insects form a greater part of biodiversity. Considering this vast diversity of insects, there is a need to explore the diversity of the insects. Taxonomy being a part of systematic presents a huge potential to study the systematics.

In the current scenario there is not even a single book which deals with all the competitive aspects of insect taxonomy for the students.

The book entitled Descriptive and Objective Insect Taxonomy is comprehensive treatment of text book cum objective type questions on all aspect of insect taxonomy. This book covers the entire field of insect taxonomy step by step in the form of notes / objective questions and answers. This book will be helpful in recollecting knowledge within a short period of time and meet the requirement of students those who are appearing in different competitive examinations such as ICAR-JRF, ARS /NET, SRF, etc.and also useful to B.Sc (Ag.)/M.Sc (Ag) Agricultural Entomology students.

It is my great pleasure to express my sincere thanks to the publisher, NIPA, New Delhi for bringing this book in a systemic way within a period of time.

We hope this book will serve as a useful guide for proper knowledge of insect taxonomy. Suggestions for future improvement of this book are most welcome and shall be highly appreciated.

Anil Kumar

Contents

Preface .... *v*
Terminology .... *ix*

1. Introduction 1
- 1.1 Taxonomy 1
- 1.2 Arthropoda 29
- 1.3 Hexapoda 30

2. Entognatha 35
- 2.1 Protura (Pro-first, ura-tail) 35
- 2.2 Diplura (Di-two, plura-tails) 36
- 2.3 Collembola (Colle-glue, emblos-piston or peg) 36

3. Ectoganatha (Apterygota) 39
- 3.1 Archaeognatha (Archaeoes-ancient/old, gnatha-jaw) 39
- 3.2 Thysanura (Thysan-tassel/bristle/fringed, ura-tail) 40

4. Palaeoptera 41
- 4.1 Ephemeroptera (Ephemero-a day or short lived, Ptera-wing) 41
- 4.2 Odonata (Odontos-tooth) 42

5. Oligoneoptera 45
- 5.1 Plecoptera (Pleco-folded/pleated/ twisted, ptera wing) 45
- 5.2 Blattodea (Blatta light avoiding insect) 46
- 5.3 Isoptera (Iso-equal, ptera-wing) 47
- 5.4 Mantodea (Mantodea mantis) 49
- 5.5 Grylloblattodea (Gryll-cricket and blatta-cockroach) 50
- 5.6 Dermaptera (Derma-skin, ptera-wings) 51
- 5.7 Orthoptera (Ortho-straight; Ptera-wings) 51
- 5.8 Phasmotodea (Phasma- phantom or apparition) 54
- 5.9 Mantophasmatodea (Combination of the order names for Mantodea and Phasmatodea) 55

5.10. Embioptera (Embio-lively, Ptera-wing) 56

5.11. Zoraptera (Zor-pure, Aptera-wingless) 56

6. Paraneoptera 59

6.1. Order: Psocoptera (Psokos-Rubbed or gnawer; ptera-wing) 59

6.2. Order: Phthiraptera (Phthir-lice, aptera-wingless) 60

6.3. Hemiptera (Hemipterus-half-winged) 61

6.4 Thysanoptera (Physopoda) (Thysano-Fringed / Physopoda-bladder footed and pteron-wing) 69

7. Endopterygota (Holometabola) 73

7.1. Order: Neuroptera (Neuro-nerve and Ptera-wing) 73

7.2. Order: Raphidioptera (Raphid-needle, and ptera-wings) 75

7.3. Order: Megaloptera (mega-large and Ptera-wing) 75

7.4. Order: Coleoptera (Coleo-sheath, pteron-wing) 76

7.5. Order Trichoptera (Trico-hair;ptera-wing) 83

7.6. Order Lepidoptera (Lepido-Scales; Pteron - wing) 84

7.7. Order: Siphonaptera (Siphon-tube like;aptera-wingles) 92

7.8. Order: Strepsiptera (Twisted-wing parasites) 93

7.9. Order: Mecoptera (Meco - long and ptera- wings) 93

7.10. Order: Diptera (Di –Twice; pteron- wings) 94

7.11. Order: Hymenoptera (Hymen- Membranous; pteron -wing) 99

8. Collection and Preservation of Insects 105

9. Objective Question 115

9.1 Multiple Choice Questions 115

9.2 True/ False 131

9.3 Matching 134

9.4 Fill up the blanks 139

9.5. Quiz 144

9.6 Special organ/character 161

Terminology

1. Acari: The order of arthropods within the class Arachnida that comprises the mites and ticks.
2. Aculate: With aculae (Lepidoptera) and with a sting(hymenoptera).
3. Adecticus: Atype of pupa in wich the mandible are immovable and non functional.
4. Alate: Those insects are possessing wings.
5. Alpha taxonomy: The level of taxonomy concerned with the characterization and naming of species. In alpha taxonomy, a grade refers to a taxon united by a level of morphological and/or physiological complexity.
6. Alloparalectotype: A specimen from the original material, of another sex than the holotype, and designated later than the original publication of the species.
7. Allopatric speciation: Speciation dependent on geographic barriers to maintain reproductive isolation,species formation during geographical isolation .
8. Allotype: A paratype of opposite sex to the holotype .
9. Ametabola: Primitive wingless insects that do not undergo metamorphosis. Comprises the orders Thysanura, Diplura, Protura and Collembola.
10. Anal ring: An elevated ring like structure surrounding the anus in coccids.
11. Anamorphosis: Type of metamorphosis in which the juvenile stages present fewer abdominal segments than the adults; in Protura, the addition of 3 abdominal segments after hatching of the first instar from the egg.
12. Anisoptera: Suborder of odonata, including the dragonflies in which the forewings and hind wings of the adult are dissimilar in venation and usually in shape and in which the nymphs are rather stout and lack caudal gills.
13. Anoplura: An alternative name for the insect order Siphunculata which comprises the sucking lice.
14. Aphaniptera: An alternative name for the insect order Siphonaptera which comprises the fleas.
15. Apterous: Insect without wings.
16. Apterygota: An alternative term to Ametabola.
17. Arachnida: A class of arthropods closely related to insects which include spiders, mites, tick and scorpions.
18. Araneae: The order of arthropods within the class Arachnida that comprises the spiders.

19. Archaeognatha: Primitively wingless order Insecta, differing from the zygentoma by possessing monocondylar mandibles, well-developed compound eyes which meet dorsomedially; large maxillary palps with 7 segments; and an appendix dorsalis that is longer than the cerci .
20. Arthropoda: The phylum of animals to which insects and other groups belong and characterized by the possession of an external jointed skeleton, segmented body and jointed limbs.
21. Autotomy: Loss of appendages in arthropods by reflex sloughing, in phasmida often associated with problems during ecdysis, and in some species also used defensively
22. Beta taxonomy: natural classification,
23. Bionomial system: The system of dual latin names for living organism devised by the Swedish naturalist Linnaeus and now almost universally applied to living organisms.
24. Biotype: A genetically distinct strain or sub-group of a species distinguished by some behavioral or physiological difference but indistinguishable morphologically. Usually applied to strains of insect species which can overcome resistant plants.
25. Brachypterous: Insect having short underdeveloped wings that do not cover the abdomen and are generallyincapable of flight. Ex. Brown plant hopper and some grasshopper
26. Brood canal: In Strepsiptera, passage between cuticle and last larval cuticle into which open genital canals
27. Brunner's organ: In grasshoppers (Orthoptera), a soft tubercle or process located proximal in the ventral sulcus of the caudal femora, and against which, when in the usual inactive position, the caudal tibiae are closely pressed
28. Brush: footed butterflies: the Nymphalidae (Lepidoptera), so-called because the fore tarsi are short and clothed with long hairs, especially in males.
29. Caddisfly: A member of the order Trichoptera.
30. Caelifera: Suborder of Orthoptera, including the superfamilies Acridoidea, Tetrigoidea, and Tridactyloidea, the members of which have antennae with fewer than 30 segments.
31. Campodeiform: A form of insect larva with well-developed legs, prominent antennae and usually several tail filaments e.g., many larval coleoptera, trichoptera, and planipennia.
32. Capitulum: A small head like structure in certain mites and ticks on which the mouthparts are borne also known as false head.
33. Caste: Structurally and functionally distinct group within the colony of a social insect eg. Worker caste.
34. Caterpillar: A polypod or cruciform larva of a moth, butterfly, or sawfly.
35. Caudal gills: In nymphal Zygoptera (Odonata), the 3 foliaceous caudal external tracheal gills.

36. Cerci: One of pair of appendages at the tip of the abdomen of insects.
37. Chelicerae: The inner portions of the mouthparts of mites; modified to piercing stylets in plant feeding mites.
38. Chilopoda: The class of arthropods that comprises the centipedes.
39. Chrysalis: The pupal stage of butterflies.
40. Chrysomeloidea: Superfamily within the Polyphaga (Coleoptera), including the families chrysomelidae, cerambycidae, and bruchidae, adults possessing 5-5-5 pseudotetramerous tarsi (penultimate tarsomere very small and hidden in the emargination of the 3rd tarsomere), a head that is not produced into a rostrum, 2 gular sutures, metasternum with a transverse suture, antennae never elbowed, abdomen with 5 visible sternites, and hind coxae without a declivity or cavity into which the femora can retract.
41. Cladistics: School of symstematics that seeks to recognize monophyletic groups on the basis of shared derived characters.
42. Cladogram: A branching diagram that represents the distribution of derived characters states among taxa,by extension, the claogram then represents the relationship among those taxa and their evolutionary history.
43. Class: A division of the animal kingdom below a phylum but above an order eg, the class Insecta.
44. Coarctate: Pupal form in which the last larval skin in retained as an extra covering; occurs particularly in higher Diptera.
45. Cocoon: A silken covering produced by some insects which enclosed and protects the pupa.
46. Coleoptera: The order of insects comprising the beetles & weevils.
47. Collembolan: The order of insects comprising the springtails.
48. Collophore: A tuve like structure on ventral side of the 1st abdominal segment of most collembolan.
49. Complete metamorphosis: Metamorphosis in which the insect proceeds through four distinct development stages: egg, larva, pupa and adult.
50. Corbicula: A smooth area on the outer surface of the hind tibia,bordered on each side by a fringe of long,curved hairs,which serve as pollen basket in bee.
51. Corium: The elongared usually thickrned,basal portion of front wing in hemiptera.
52. Chorotype: A fossil specimen collected from the same stratum as the type, but from a neighboring locality.
53. Cornicle: A tubular process (paired) which occurs on the abdomen of aphids
54. Crawler: The first instar nymph of scale insects which possesses legs and thus enables dispersal to take place.
55. Crochets: Hooked spines at the tip of the prologs of lepidotera larvae.
56. Crustacea: A class of arthropods, mostly aquatic in habit, to which crabs and cray fish belong.

57. Cryptic species: Sexually isolated populations with few or no tangible recognition characters to set them apart from the general species population .
58. Cytotaxonomy: Use of cytology and cytogenetics in elucidating the relationships of higher taxa and comparison of the chromosome sets of related species.
59. Dermaptera: The order of insects comprising the earwigs.
60. Deuteronymph: The third instar stage in the development of mites and ticks.
61. Decticous: A type of pupae with movable,functional mandible.
62. Dictyoptera: The order of insect comprising the cockroaches and mantids.
63. Dichotomous: Forked divided or dividing into 2 parts of taxonomic keys, arranged into couplets.
64. Dichotomy: A branching of a single stem into 2 equal and diverging branches, used in phylogeny of ancestral lines of descent, and in venation of main veins or their branches; also, a table or key for determining species or higher groups, in which they are separated by contrasting characters arranged in couplets, 2 by 2.
65. Diplopoda: The order or arthropods within the class Myriapoda that comprises the millipedes.
66. Diplura: The order of insects that comprises the two-pronged bristle tails.
67. Diptera: The order of insects that comprises the true flies which are characterised by possession of only one pair of wings.
68. Drone: The male of social bees and wasps.
69. Dun: Alternative term of sub-imago have winged but sexually immature mayfly adult which has to undergo another moult.
70. Ectoparasite: A parasite which lives externally on the body of its host.
71. Elytra: The hard horny forewings of beetles' i.e Coleoptera which act as protective cases for hind wings.
72. Emendation: In taxonomy, an intentional change to a previously proposed name, e.g., Lindinger proposed the emendation Hemiberlesea for the armored scale Hemiberlesia indicating that it was originally improperly formed.
73. Embiopterara: Small order of tropical insects which live gregariously in silken nests as web spinners.
74. Endoparasite: A parasite which lives internally within the body of its host.
75. Endopterygota: Holometabola, undergo complete metamorphosis with larva, pupa and adult stages.
76. Ephemeroptea: The order of insects comprising the mayflies.
77. Epidemic typhus fever: A human disease caused by Rickettsia prowazekii and transmitted by the body louse, *Pediculus humanus humanus*.
78. Evolutionary classification: A form of natural classification which recognizes both paraphyletic and monophyletic taxa.

79. Exarate: A pupae having legs and other appendages separately encased as in the pupae of beetles, bees, and wasps.
80. Exopterygota: Hemimetabola that undergo incomplete metamorphosis.
81. Family: The main unit of classification into which genera are grouped and insect family names always ending with- idae.
82. Furca: Aforked structure that support in the upper part of the mouthparts of flies.
83. Gradual metamorphosis: Alternative term for incomplete or partial metamorphosis.
84. Gregarious: The habit of living together in groups.
85. Grub: A scarabaeiform larva which is thick bodies with thoracic legs and well developed head and usually sluggish.
86. Grylloblattodea: A small order or primitive insects related to the cockroaches and crickets for which there is no common name.
87. Haltere: The knob shaped modified hind wings of Diptera which act as balancing organs.
88. Hemelytron: The forewing of Hemiptera in which the basal portion is thickened and the distal portion membranous.
89. Hemimetabola: Alternative term of Exopterygota.
90. Hemiptera: The order of insects that includes cicadas, leafhoppers, scale insects, aphids, plant bugs an many other groups and characterised by the possession of piercing and sucking mouthparts.
91. Heteroptera: The Heteroptera includes the true bugs.
92. Holotype: A single specimen designated as the name-bearing type of a species or subspecies when it was established, or the single specimen on which the taxon was based when no type was specified .
93. Holometabola: Alternative term of Endopterygotay.
94. Homonym : One of two or more scientific names that are identical but pertain to different organisms.
95. Homoptera: A sub-order of the insect orders Hemiptera and includes cicadas, leathoppers, aphids and scale insects.
96. Hymenoptera: The order of insects that includes saw flies, ants, bees and wasps.
97. Imago: The adult sexually mature form of an insect.
98. Isoptera: Order of insects comprising the termites.
99. Lectotype: A syntype designated as the single name-bearing type specimen subsequent to the establishment of a nominal species or subspecies.
100. Lepidoptera: Insects comprising the butterflies and moths
101. Maggot: A vermiform larva without legs and without well-developed head capsule.

102. Mallophaga: The order of insects comprising the biting or bird lice.
103. Mecoptera: A small order of insects comprising the scorpion flies.
104. Mite: A member of the class Acari related to but distinct from the insects. Mites may be distinguished from insects by the single body region, four pairs of legs and absence of antennae.
105. Mollusca: The phylum of animals to which slugs and snails and their relatives belong.
106. Monograph: In taxonomy, an exhaustive treatment of a taxon, including all information pertinent to taxonomic interpretation, comparative anatomy, biology, ecology, and distribution.
107. Myriapoda: Class of arthropods which includes millipedes and centipedes.
108. Neotype: The single specimen designated as the name-bearing type of a nominal species or subspecies for which no holotype, or lectotype, or syntype(s), or prior neotype, is believed to exist.
109. Neuroptera: A order of insects that includes lacewings and ant lions.
100. Nits: The eggs of a louse.
111. Odonata: The order of insects comprising the dragonflies & damselflies.
112. Ootheca: An egg case or egg capsule of distinct form as produced by cockroaches.
113. Opisthosoma: The hind body region of arachnids includes spiders, mites, ticks.
114. Order: A division of a class or sub class comprising a group of related family.
115. Orthoptera: The order of insects that includes crickets, grasshoppers, locusts and wetas.
116. Paleontology: The study of ancient life, on the basis of fossil or other remains.
117. Phasmida: Order of insects comprising stick and leaf insects.
118. Phylum: A major division of the animal kingdom corresponding to the main animal types, e.g., phylum: Arthropoda-arthropods, phylum: Mollusca: molluscs.
119. Physogastry: Enlargement of the abdomen of female due to increased size of reproductive system, e.g., termites.
120. Plecoptera: The order of insects comprising the stoneflies.
121. Prosoma: The foremost body region of arachnids spiders, mites, ticks to which the walking legs are attached.
121. Protonymph: The second instar in the development of mites and ticks.
122. Pseudopods: False legs on the abdomen of the larvae of butterflies, moths and sawflies.
123. Psocoptera: The order of insects comprising the book lice and bark lice.
124. Ptilinurn: In Diptera an organ that can be inflated to a bladder-like structure and thrust out through frontal suture of the head at the time of emergence from the puparium.

125. Puparium: Specialised form of pupa in which the last larval skin is retained as an extra covering and characteristic of the higher Diptera
126. Queen: Female sexually reproductive form of social insects such as bees.
127. Taxonomy: The study of the naming and classification of living.
128. Tegmina: The leathery forewings of insects such as grasshoppers or cockroaches.
129. Termitaria: The nests build by termites under or above the ground.
130. Thysanoptera: The order of insects comprising the thrips.
131. Thysanura: The order of insects that includes three-pronged bristle tails such as silverfish.
132. Tick: A member of the order Acari within the class Arachnida;
133. Topotype: One or more specimens collected at the same location as the type series regardless of whether they are part of the type series.
134. Tribe: A unit of classification below the family level into which groups of like genera are placed and names always ends with-ini.
135. Trichoptera: The order of insects comprising the caddisflies.
136. Ventral tube: A tubular protrusion from beneath the first abdominal segment of springtails.
137. Vertical classification: It is described by Simpson that a taxon based on ancestor and descendant (phylogenetic) relationship between its members, a clade. Evolutionary systematics considers both horizontal and vertical classification in taxonomy, whereas cladistics and phylogenetics is based on vertical classification only.
138. Workers: Sterile individuals in the nests of social insects which forage for food and generally run the colony.
139. Zoraptera: A small order of tropical insects related to the Angel insects.

1

Introduction

Insects are the most successful arthropods in animal kingdom. They are found in almost all habitats on earth except for the frozen polar environments at the highest altitudes and in the immediate vicinity of active volcanoes. They are the only invertebrate animal (animals without backbones) with wings. The success of insects on the planet is due to their ability to fly and rapidly colonize new habitats. The study of insects is called entomology and entomologists are those scientists who study about insects.

The word entomology is derived from Greek word, *entomon* = insect and *logos* = study. It is a branch of zoology which deals with the scientific study of insects. Insects constitute the largest Class of the whole living organisms and about 70-75 per cent of all living animals are insects with 10-15 lakh known species and many are yet to be discovered.

Insects play a vital role in the ecosystem. Some insects are pollinators of flowering plants, beside being a source of food for insectivorous animals and assisting in the decomposition of plants and animals. It is an ancient group of animals and first probably appeared before the Devonian period beween 400-360 million years ago (mya) and by the Carboniferous period (360-285 mya) flight adaptation appeared. Adaptation to flight proved a highly successful strategy and during the Permian period (285-245 mya) insect achieved their greatest diversity. No other arthropod group has developed capability to fly during the process of evolution. By the Permian period, the basic physical structure of many of the modern orders of insects had evolved. The more recently evolved Hymenoptera (ants, bees, wasps and sawflies) and Lepidoptera (butterflies and moths) appear as fossils in the Jurassic period (210-145 mya). The Mantodea (praying mantids) appeared in Eocene epoch of Paleogene period in fossilised amber (60-35 mya).

1.1 Taxonomy

Taxonomy word is derived from Greek word *(Taxis*- arrangement; *nomos*-law)

- Word taxonomy was first proposed by AP de Candolle in 1813
- Carl Linnaeus is considered as 'Father of taxonomy'

Taxonomy: It is process of identifying and classifying living organism. Simpson (1961) defined taxonomy as the theoretical study of classification, including its basis, principles, procedures and rules.

Other definition of taxonomy

- Mason (1950) defined taxonomy as the synthesis of all the facts about the organisms into a concept and expression of the interrelationship of organisms.
- Harrison (1959) defined taxonomy as the study of principles and practices of classification, in particular, in methods, the principles and even in part, the result of biological classification.
- Heywoods (1967) defined taxonomy as the way of arranging and interpreting information.
- Blackwelder (1967) explains it as the day to day practice of handling different kinds of organisms. It includes collection and identification of specimens, the publication of data, the study of literature and the analysis of variations shown by the specimens.
- Johnson (1979), taxonomy is the science of placing biological form in order.
- Christoffersen (1995) defines taxonomy as the practice of recognizing, naming and ordering taxa into a system of words consistent with any kind of relationships among taxa that the investigator has discovered in nature.

Why Taxonomy?

- Classification - Arrangements in groups (taxa)
- Nomenclature - Assigning names to taxa
- Identification - Determination of taxon to which an isolate belongs.

STAGES OF TAXONOMY

- **Alpha taxonomy**: (Analytical phase) The level of taxonomy concerned with characterization and naming of a species on the basis of morphology.
- **Beta taxonomy**: (Synthetic phase) The level of taxonomy concerned with the arrangement of a species in the biological hierarchy.
- **Gamma taxonomy**: (Biological Phase) The level of taxonomy concerned with the studies on intra- specific population to understand speciation, evolutionary rates and trends.

Some important developments & Contributions of Scientist in the field of taxonomy:

- Linnaeus (1735) - Natural classification
- Darwin (1859) - Phylogenetic classification

- Carl Woese (1977) - Modern Phylogeny
- Cyber taxonomy (New field) termed coined by Wheeler in 2007. Cyber taxonomy is the contraction of "cyber enabled taxonomy" with the use of modern cyber infrastructure and adoption of digital technology, cybertaxonomy is able to produce result at a very faster rate.

Systematic: It can be defined as the science of the diversity of organism and the relationships among them.

- Systematics is derived from Latin word '*systema*', which means systematic arrangement of organisms. Carl Linnaeus used 'Systema Naturae' as the title of his book.
- Relationships are visualized as evolutionary trees (synonyms: cladograms, phylogenetic trees: phylogenies).
- Systematists are also vitally interested in determining the evolutionary history of species and the features that result in adaptation to the environment.

Phylogenies have two components

- Branching order (showing group relationships) and branch length (showing amount of evolution).
- Phylogenetic trees of species and higher taxa are used to study the evolution of traits (e.g., anatomical or molecular characteristics) and the distribution of organisms (biogeography).

Other definition of systematic

- According to Blackwelder (1967) , systematics is that science which includes both taxonomy and classification, and all other aspects of dealing with kinds of organisms and the data accumulated about them.
- Christofferson (1995) defined systematic as the theory, principles and practice of identifying (discovering) systems, i.e., of ordering the diversity of organism (parts) into more general systems of taxa (wholes) according to the most general causal processes.
- According to Blackwelder and Boyden (1952) , "systematic is the entire field dealing with the kind of animals, their distinctions, classification, and evolution.
- Simpson (1961) defines systematic as "The scientific study which deals with kinds and diversity of organisms and any or all relationships among them'.

History of Systematics

- Aristotle (384-322 BC, father of zoology) arranged animals on the basis of habitat into aquatic, terrestrial, aerial animals.
- On the basis of single character, Greek scholars divided animals into four major groups – insects, birds, fishes and whales.
- Pliny the Elder (28-79 A.D.) introduced the first system of artificial classification.
- Swedish naturalist, Carolus Linnaeus developed the scientific system of naming species. It is known as binomial system of nomenclature.

Importance of Systematics

- Systematics plays a central role in biology by providing the means for characterizing the organisms that we study.
- By allowing taxa (taxonomic groups) to be correctly identified classifications provide a key to the literature and a means for organizing information.
- Pest management entomologists must always keep in mind that systematics, like all scientific endeavors, is not a static field.
- In many cases, identifications are required and made before a group has been adequately studied taxonomically.

ROLE OF SYSTEMATICS

- To study diversity of organism
- It plays a crucial role in applied biology
- It also helps in environment protection
- Finding new species
- Biodiversity conservation
- Documentation
- Pest management

Characters and their use in Systematics

- Morphological
- Physiological
- Molecular
- Ecological
- Reproductive and
- Behavioral

Basics of Systematic Study

- **Characterization:** The organism to be studied is described for all its morphological and other characteristics.
- **Identification:** Based on the studied characteristics, the identification of the organism is carried out to know whether it is similar to any of the known group or taxa.
- **Classification:** The organism is now classified on the basis of its resemblance to different taxa. It is possible that the organism may not resemble any known taxa or groups. A new group or taxon is raised to accommodate it.
- **Nomenclature:** After placing the organism in various taxa, its correct name is determined. If the organism is new to systematics, it is given a new name based on rules and conventions of nomenclature.

Function of insect systematic

- To organize our knowledge about the organism.
- To provide means of communication.
- To provide framework of classification and identification of organism.
- Universal and stable nomenclature
- Phylogenic relationship

Principles and methods of systematic

- Phylogenetics which determines the relationships of natural taxa based on their evolutionary history.
- A natural taxon is a group of organisms that are a result of evolution.
- Natural taxa are called monophyletic groups (or clades) which is a group of species that includes the most common ancestor.
- All of its descendants artificial groups are those that do not confirm to evolutionary processes or history (polyphyletic and paraphyletic groups).
- Phylogenies are expressed as tree like diagrams, and represent the genealogic relationships of the study taxa.

Relationship of taxonomy to the systematics

Kapoor (1998) - considered that the relationship of taxonomy to systematics is somewhat like that of theoretical physics to the whole field of physics. Taxonomy includes classification and nomenclature but systematics includes both taxonomy and evolution. In simple terms, actually there are two parts of systematic.

The major differences between taxonomy and systematic can be summarized as follows:

- Taxonomy is the most important branch of systematic and thus systematics is a broader area than taxonomy.
- Taxonomy is concerned with identification, description, classification and nomenclature of a species, but systematics is important to provide layout for all those taxonomic functions.
- Evolutionary history of a species is studied under systematics but not in taxonomy.
- The environmental factors are directly related with systematics but in taxonomy it is indirectly related.
- Taxonomy is subjected to change in course of time, but systematics is not changed if it was properly done.

Classical Taxonomy and Modern Taxonomy

Classical Taxonomy- The classical taxonomy is based on observable morphological characters with normal individuals considered to be expression of the same while their variations are believed to be imperfect expressions. Classical taxonomy originated with Plato followed by Aristotle (Father of Zoology), Theophrastus (Father of Botany) up to Linnaeus (Father of Taxonomy) and his contemporaries.

- Species are delimited on the basis of morphological characteristics.
- Only a few characters are employed for classification.
- A few individuals or their preserved specimens are used for study. It is called typological concept.
- Species are believed to be static.
- Species is the centre stage of study. Its subunits are not important.

Modern Taxonomy/New Systematics

- The term new systematic was coined by Julian Huxley in 1940.
- New systematic is systematic study which takes into consideration all types of characters.
- Besides classical morphology, it includes anatomy, cytology, physiology, biochemistry, ecology, genetics, embryology, behavior etc. of the whole population instead of a few typological specimens.
- New systematics is also called population systematics and biosystematics.

- It strives to bring out evolutionary relationship amongst organisms.
- New systematic is based on the all types of variation in the species.
- Along with morphological variations, other investigations are also carried out to know the variety of traits.
- Delimitation of species is carried out on the basis of all types of biological traits. It is also called biological delimitation.
- Traits indicating primitiveness and advancement are found out.
- Inter-relationships are brought out.
- Species are considered as dynamic unit.

Taxonomic key

- Taxonomic key is a device for easily and quickly identify to unknown animal where belongs.
- The key consists of a series of choices, based on observed features of the specimen.
- It provides a choice between two contradictory statements resulting in the acceptance of one and the rejection of the other.
- A single pair of contradictory statements is called a couplet and each statement of a couplet is termed a lead.
- By making the correct choice at each level of the key, one can eventually arrive at the name of the unknown animal.

Two types of keys

- Dichotomous
- Poly clave

Dichotomous Keys

- Keys in which the choices allow only two (mutually exclusive) alternative couplets are known as dichotomous keys.
- Dichotomous comes from the Greek root *dich* meaning "two" and *temnein* meaning "to cut".
- The couplets can be presented using numbers (numeric) or using letters (alphabetical).
- The couplets can be presented together or grouped by relationships.
- There is no apparent uniformity in presentation of dichotomous keys.

Two types of dichotomous keys

- **Indented Keys (also called yoked):**
- Indents the choices (leads) of the couplet an equal distance from the left margin.
- The two choices of the couplet are usually labelled 1 and 12 or la and lb.
- It is not necessary that the choices are numbered, but it helps.
- The user goes to the next indented couplet following the lead that was selected.
- **Bracketed Keys:**
- Provides both choices side-by-side.
- The choices of the couplet must be numbered (or lettered) .
- It is very helpful if the previous couplet is given.
- This key has exactly the same choices as the first example.
- The choices are separated, but it is easy to see the relationships.
- While this key might be more difficult to construct, it gives more information to the user.

Poly Clave Keys (also called Multiple Access or Synoptic)

- Another type of key, which is relatively a new alternative to dichotomous keys and becoming increasingly popular, especially because of the ease of computerizing them, is termed multiple access or poly clave or synoptic key.
- The advantage of these keys is that they allow the user to enter the key at any point.
- This key is based on the identification of organisms by a process of elimination. In a written poly clave key, there is a series of characters and character states. Each state is followed by a number or code for the species that possess that feature.
- The user needs to select any character and then copy down the list of species that possess the feature. Then the user has to select another character and eliminate any species that is not common tò both lists. This process has to be continued until the specimen is identified.

Advantages

- Easy to use.
- They allow multi-entry i.e.; the user can start anywhere.

- They are order-free i.e., the user can work in any direction with any character.
- They are faster.
- They are easily computerized.

Analogy vs homology

Homology	Analogy
In biology, homology is the resemblance of the arrangement, physiology, or growth of various species of organisms.	In biology, an analogy is a functional similarity of structure, based on the similarity of use and not upon common evolutionary origins.
They do not have similar functions	They have similar functions.
Homologous structures are inherited from a common precursor.	Analogous structures are inherited from a different precursor.
Homologous structures are developed from related embryonic substances.	Analogy is developed in species that are not related to each other.
Homology is the result of divergent evolution.	Analogy is the result of convergent evolution
Example: The human arm, the wing of the eagle, and the pectoral fins of the whale are homologous structures, as they all have identical patterns of bones, muscles, nerves, and blood vessels with similar developmental origins. But they have a different purpose, though.	Example: The wings of birds and insects, on the other hand, are equivalent structures. They both permit flight but do not have any developmental processes in common.

Similarities between Parallel and Convergent Evolution

- Divergent, Parallel and convergent evolution are three types of evolutionary patterns.
- Parallel and convergent evolution occur independently in distinct species.
- Both parallel and convergent evolution occur under the influence of same environmental pressures.
- Both parallel and convergent evolution do not lead to speciation.

Difference between Parallel and Convergent Evolution

- Parallel evolution refers to the independent evolution of similar traits in different but equivalent habitats but convergent evolution refers to the independent evolution of analogous structures in unrelated species in same habitat.
- Parallel Evolution has two distinct species evolve independently; maintaining the same level of similarity but Convergent Evolution has two distinct species evolve analogous traits.
- Parallel evolution occurs in unrelated or distantly-related species but convergent evolution occurs in unrelated species.

- Examples of Parallel Evolution are in butterflies, having similar wing colouration and patterns, both within and among families. Whereas example of Convergent Evolution is development of the eye of vertebrates and arthropods.

Monothetic and Polythetic taxa

Monothetic taxa: When presence or absence of a trait is considered for inclusion of a species in a particular taxon.

Polythetic taxa: Polythetic systems allow a taxon to be defined on the basis of a greater number of mutually interchangeable traits. Modern systems are mostly polytheic. As compared to monothetic systems, polythetic systems are somewhat harder to use and its classification schemes i.e. as instruments for determination of organisms. On the other hand, only polythetic systems can be truly natural, i.e. can reflect the phylogenesis of the studied organisms.

Paraphyletic taxa: A group composed of a collection of organisms, including the most recent common ancestor of all those organisms. Unlike a monophyletic group, a paraphyletic taxon does not include all the descendants of the most recent common ancestor.

Sexual dimorphism

Sexual dimorphism is the phenomenon where the opposite sexes of the same species exhibit different phenotypic characters, such as secondary sexual characteristics, size, colouration etc. especially which are not directly involved in reproduction.

Insects present a wide veriety of sexual dimorphism such as:

- Males of the beetle *Oedemera nobilis* have swollen femurs whereas the females have slender femurs.
- Females of fireflies do not have a functional wing.

Classification of animals

1. Artificial classification adopted by Pliny the Elder on the basis of morphological character
2. Natural classification proposed by Bentham and Hooker on the basis of anatomical, physiological, pathological, biochemical, reproductive and cytological character.
3. Phylogenetic system of clasifiation: first given by Engler and Prantl on the basis of evolutionary aspect of organism.

Some important facts

- Aristotle is considered as the 'Father of biological Classification
- 1st classification develop by Plato and Aristotle
- The classification of animals was first started by Aristotle
- Later by Linnaeus who is considered as the father of taxonomy and classified insects into only seven orders *viz* Coleoptera, Hemiptera, Lepidoptera, Neuroptera, Diptera, Hymenoptera and Aptera.
- Jeannel recognized class into 40 orders
- Brues, Melander and carpenter recognized 27 orders
- Imms and pruthi gave 29 orders
- Essig and Mani listed 33 orders
- Ross 28 orders
- The Linnaeus Method, also known as Linnaean Taxonomy, creates a hierarchy of grouping called taxa, as well as binomial nomenclature that gives each animal species a two-word scientific name.
- This method of giving scientific names to animals is typically rooted in Latin by combining the genus and species.

SCHOOLS OF CLASSIFICATION

PHENETIC, EVOLUTIONARY AND CLADISTIC CLASSIFICATION

Cladistic or phylogenetic classifications

- Cladistic classifications only include monophyletic groups (common ancestry) because only they have the unambiguous hierarchical arrangement of the phylogenetic tree. This is the theoretical beauty of cladism.
- Only monophyletic groups are formed in the cladistic conversion of a phylogenetic tree into a classification and nothing has to be known about the phenetic evolution of species within each branch. It believes in cladogenesis, where two taxa originated in the same branching event have a common ancestor that is not shared by any other taxon.

Phenetic or Numerical phenetics

- Numerical phenetics attempts to measure similarity using observable attributes of organisms (phenotypic characters). The school emphasizes quantitative techniques of measurement.

- Peter H.A. Sneath and Robert Sokal define numerical taxonomy as "the grouping by numerical methods of taxonomic units into taxa on the basis of their characters states."
- The simplest kind of numerical phenetic classification is defined by only one or two characters.
- The best phenetic classification is one constructed by considering as many attributes as possible.
- Numerical pheneticists therefore recommend measuring as many characters as possible - even hundreds - and classifying according to the aggregate similarity for all of them. The more characters that are measured, the more likely it is that peculiar individual characters will be averaged out, and the better founded and more natural the classification will be. Hence, the development of computers have made possible to handle a huge set of attributes
- The trouble with this procedure is that different individual characters show different distributions among species and therefore tend to produce different classifications.

Evolutionary classification

- Evolutionary classification is a synthesis of the phenetic and phylogenetic principles.
- The school therefore describes itself as synthetic, drawing on the advantages, and avoiding the shortcomings, of the two purer schools of cladism and numerical phenetics.
- However, for the same reason it has been criticized for doing the opposite - for retaining the philosophical shortcomings of phenetic classification and adding to them the practical uncertainties of phylogenetic inference.
- Evolutionary classification permits paraphyletic groups (which are allowed in phenetic but not in cladistic classification) and monophyletic groups (which are allowed in both cladistic and phenetic classification) .
- Since it defines groups by homologies and ignores homoplasies it excludes polyphyletic groups (which are banned from cladistic classification but permitted in phenetic classification) .
- The school of evolutionary classification was developed by the biologist Ernst Mayr.

Zoogeography: Geographic distribution of animals past and present.

To find out the current distribution of taxa and how these patterns relate to the evolutionary history.

Alfred Russel Wallace is the father of zoogeography. He was the first to realize that distribution of related species is fundamentally linked ton geologic history of places the species inhabit.

Two hypothesis has been proposed for the formation of zoogeographic regions:

Dispersal: A small population is dispersed to a different geographic region and it forms peripatric species.

Vicariance: the geographical separation of a population, typically by a physical barrier such as a mountain range or river, resulting in a pair of closely related species. Hence, it forms allopatric species.

A monophyletic group is a group which contains all the descendants of a common ancestor: the group has a common ancestor unique to itself. It is the only group recognized by cladist classification. Monophyletic groups contain all the branches below a given ancestor; nothing is said about the phenetic evolution of species within each branch.

A paraphyletic group contains some, but not all, of the descendants from a common ancestor. The members included are those that have changed little from the ancestral state; those that have changed more are excluded: a paraphyletic group contains the rump of conservative descendants from an ancestral species.

Polyphyletic groups are formed when two lineages convergently evolve similar character states. Organisms classified into the same polyphyletic group share phenetic homoplasy's as opposed to homologies.

Cladogram

Cladogram: A cladogram is diagrammatic representation of an evolutionary tree that shows the ancestral relationships among organisms.

A cladogram have the following component:

Root: it is the starting point in any cladogram. However, a root might have originated from larger clade.

Node: A node is a hypothetical ancestor which through speciation gave rise to different daughter taxa. It indicates the bifurcating branch point of divergence in all cladograms.

Clade: A branch that includes a single common ancestor and all of its descendants is called a clade.

Phenogram

It is a diagram depicting taxonomic relationships among organisms based on overall similarity of phenetic characters without regard to evolutionary history.

Difference between cladogram and phenogram

Cladograms rely on ancestral assumptions while phenogram don't take evolutionary history into account.

Branch lengths provide no data in cladogram but are very important in phenograms.

Cladograms emphasize more on ancestral relationships and phenograms emphasize more on current relationships.

IMPORTANCE OF CLASSIFICATION

- It makes easy to study the extensive form of organisms.
- It helps to recognize the interrelationships amongst exclusive organisms.

Basis of Classification

- Natural - anatomical characteristics
- Phenetic - phenotypic characteristics
- Genotypic - genetic characteristics
- Phylogenetic - evolutionary links

Hierarchical arrangement

- The taxonomic hierarchy or hierarchy of categories was first established by Linnaeus in 1758) in the animal kingdom. The seven major categories are: Kingdom, Phylum, Class, Order, Family, Genus, and Species.
- The classification of the animal kingdom is hierarchical and follows as sequence in descending order of rank.

PHYLUM → SUBPHYLUM → DIVISION → CLASS → COHORT → ORDER → SUBORDER → INFRAORDER → SUPERFAMILY → FAMILY → SUBFAMILY → TRIBE → SUBTRIBE GENUS → SUBGENUS → SPECIES → SUBSPECIES

Category: It is only an abstract term and it represents a rank or level in classification.

Taxon : (pl. Taxa) A taxon is a unit of any rank i.e. kingdom, phylum, class, order, family, genus, species designating an organism or a group of organisms.

The following hierarchical level endings with standardized suffix are:

Level	**Ending**
Superfamily	oidea
Family	idae
Subfamily	inae
Tribe	ini
Subtribe	ina

- Species is unit of classification and it is necessary to understand a different kind of types which is recognized by international code of zoological nomenclature was evolved in1901 at Berlin.

Species concept

The term "Species" is differently perceived by different scientists, Cuvier (1829) defined species as the assemblage descended from one another or from common parents and of those, who resemble each other. Thompson defined species as "The group of individuals distinguished by an inducible set of constant properties and connected by descent and genetic relationship". According to Huxley (1942) the species can be regarded as "a geographical definable group, whose numbers actually interbreed or are potentially capable of interbreeding in nature, which normally in nature does not inter breed freely or with full fertility with related groups and is distinguished from them by constant morphological differences".

Mayr (1953) reviewed all these concepts of species and grouped them under three broad categories:

1. Typological / Essentialist / Morphological / Phenetic species concept: Typology is based on morphology/phenotype of the organism. Based on a similar morphology of the organism different species are categorized.

Criticism: Sexual dimorphism: In sexually dimorphic species, the males and females are to be assigned different species if this concept is used.

2. Nominalistic species concept: This concept was put forward by Occam and his follower, who believed that only individuals exist, while species are man's own creation.
3. Biological species concept: Biological species concept was given by Mayr (1940), as per this concept "species are groups of inter breeding natural populations that are reproductively isolated from the other such groups."

Therefore a species have three separate functions. Members of a species form a reproductive community and the individuals of a species recognize each other as potential mates. Species is an ecological unit that interacts as a unit with other species which shares its environment.

Species is a genetic unit that consists of a large gene pool where as an individual is merely a temporary vessel holding a small portion of a gene.

Criticism of biological species concept

- This concept is applicable only to sexually reproducing species.
- It is difficult to determine how much reproductive separation is needed to declare an organism a different species.

- The definition is applicable only to current population and ignores the species status of ancestral population, ie. If the current population is truly sympatric.

4. Evolutionary / Phylogenetic species concept: Simpson (1961) defined this concept as "A species is a series of ancestor descendent populations passing through time and space independent of other populations, each of which possesses its own evolutionary tendencies and historical fate."

- It considers sexually as well as asexually reproducing organism

5. Recognition species concept: Patterson (1985) and Lambert et al. (1987) brought this concept as a replacement to biological species concept. According to this concept a species is " that most inclusive population of individual biparental organisms, which share a common fertilization system". The fertilization system includes all aspects of organism s biology that "contribute to the ultimate function of bringing about fertilization while organisms occupy its normal habitat." The organism recognises each other as mates or passive.

- **Species:** These are a group of individuals which are similar in their structure, capable of interbreeding and producing fertile off spring, but at the same time reproductively isolated from other groups.
- **Subspecies:** is an aggregate of phenotypically similar populations of a species, inhabiting a geographic subdivision of the range of a species and differing taxonomically from other populations of the species.
- **Genus** : A group of species having some definite similar characters or relationships is called a genus
- **Subfamily** is a group of allied genera to form a subfamily
- **Family** is a taxonomic category containing a single genus or a group of genera.

INFRASPECIFIC CATEGORIES

Infraspecific name is the scientific name for any taxon below the rank of species, the International Code of Zoological Nomenclature (4th edition, 1999) accepts only one rank below that of species, namely the rank of subspecies. Other groupings or infra subspecific entities do not have names regulated by the ICZN. Article 24 of the ICZN describes how infraspecific names are constructed

- **Variety:** It can be defined as the group of population which is different from the parent population. It refers to the collective variations of local, geographical and ecological population. This was the only subdivision of the species recognized by 'Linnaeus'. Now this term is not used in animal taxonomy.

- **Sub-species:** The term "Sub-species" replacement for the term "Variety" during 19th century. The sub-species can be defined as an aggregate of phenotypically similar populations of a species, inhabiting a geographic subdivision of the range of species and differing taxonomically from other populations of the species. In simple words, it is a group of population of species which are phenotypically similar but geographically isolated. Sub-species are 'Allopatric' i.e., geographically isolated and 'Allochronic' i.e., never formed at same time & at same place. Sub-species is only valid category in Zoology e.g., *Pediculus humanus capitis* (Head louse) and *Pediculus humanus carporis* (Body louse).
- **Race:** It is defined as a category representing geographical isolates i.e., a group of individuals which are phenotypically different but found as a distinct group in a particular geographical area. It is not a recognised taxonomic category.
- **Cline:** This term was coined by J.S. Huxley in 1939 for a character gradient. It is also not a recognised taxonomic term. Cline may be defined as the group of local population of a widely distributed species, which exhibit regular or gradual stepwise modification from one part of their geographic range to another. Cline originates due to reduction in gene flow between populations of a species situated at the extremities of its geographical range due to distance.
- **Deme**: A deme is a community of potentially interbreeding individuals at a given locality which share a single gene pool i.e., they are genetically similar. Deme has been used to denote the local population of a single species. It has no taxonomic importance and has only evolutionary importance.
- **Super species**: It was introduced by Mayer in 1931. It is a monophyletic group of closely related and largely or entirely allopatric species. The component species of a super species were designated as semi-species.

Two types of name exist in insect nomenclature

- **Common name** is inaccurate and not universal.
- **Scientific name** is universal and accurate.

Common versues Scientific Name

Many organisms have common name and sometimes common name used for two or more distinctly different organism. So without a specific name it's impossible to communicate about specific organism. The binomial system is the system used to name species.

Scientific Nomenclature

- It plays important role in systematics.
- Naming of organism should be *unique, distinctive, stable* and *universally accepted.*
- Carolues Linnaeus (1707 – 1778) in his 10th edition of 'Systema naturae' published in 1758 used the binomial system of nomenclature for the first time,
- The scientific name consists of two words viz. gereric name and specific epithet.
- If printed, it should be italicized. When hand written, it should be underlined.
- The first letter of generic name should be written in uppercase whereas Specific epithet in lowercase.
- In 1842, Strickland published a code of nomenclature in English Strickland code.
- Another code called 'Dall code" was evolved by the Americans in 1877.
- 1901 at Berlin, an International Code of Zoological Nomenclature was evolved.
- At the 16th session of International Congress of Zoology in Washington, the latest International Code of Zoological Nomenclature was approved. The same was revised and published in 1964
- International Code of Zoological Nomenclature regulates the naming of species.

Principles of ICZN

Principle of Binomial nomenclature

As this principle, the scientific name of a species has a combination of two names. The name of the species is composed of Generic name and Specific name.

Principle of Priority

According to this principle, the correct formal scientific name is the oldest available valid name:

- Priority is the basic principle of nomenclature.
- All zoological names have starting date of January 1, 1758.
- Priority is the guiding principle in case of disputed in case where more than one name is proposed for a taxa called **Synonyms**.
- In this case if oldest name available, it is considered as valid name otherwise rejected on other ground.

Principle of coordination

According to this principle, when a new zoological name is published, it automatically establishes all corresponding names in relevant ranks.

Principle of First Reviser

This principle is applied in case of conflicts between published names. When a conflict arises between two simultaneously published divergent names, the first subsequent author can decide which name has precedence.

Principle of Homonymy

As per this principle, the name of each taxon must be unique and must not be replicate or duplicate of any other family, group or species.

- Here identical name proposed for more than one taxon called **homonyms**.
- Here also the oldest name available is applied that particular txon where as junior name is rejected and a new name is given to that taxon.

Principle of Typification

According to this principle, each nominal taxon in the family group, genus group or species group must have a prefixed name-bearing type. This helps in determining what name it applies to.

International Code of Zoological Nomenclature

- The set of rules, system and recommendations for zoological nomenclature authorized by the "International Congress of Zoology" is called "International Code for Zoological Nomenclature" (ICZN).
- Head office of ICZN is in London.

This code consists of three main parts

1. **The Code proper:** The Code Proper includes "Preamble" followed by 90 consecutively numbered "Articles" grouped in 18 chapters. Each article is composed of one or more mandatory provisions, which are sometime accompanied by "Recommendations".
2. **Appendices:** There are three appendices, the first two have the status of recommendations and third is the constitution of the Commission.
3. Glossary: The terms used in the text are clearly defined in the "Glossary".

ORIGIN OF THE CODE

- Linnaeus was the first to mention sets of rules of nomenclature in his book "Critica Botanica" (1737) and "Philosophica Botanica" (1751).

- During 19th century more and more new and local sets of rules originated in different countries.
- **Blancherd** prepared sets of rules, which were presented at the 1st "International Congress of Zoology" held in 1889 in Paris, but accepted only in the 2nd "International Congress of Zoology", Moscow, 1892.
- Later after a revision a final draft was prepared and circulated to the "International Commission of Zoological Nomenclature", on 11th Jan,1961.
- It was approved by the Commission and formally accepted as the "International Code of Zoological Nomenclature" adopted by the "XV International Congress of Zoology", London, 1958 and published by "International Trust for Zoological Nomenclature", London 1961.
- The objective of the Code was highlighted in the first edition of the International Code of Zoological Nomenclature, 1961.
  - 1st edition - 1961
  - 2nd - 1963
  - 3rd - 1985
  - 4th - 1999 (effective from 2000)

OBJECTIVE OF THE CODE

- **Uniqueness:** Every scientific name has to be unique because it is the key to its entire literature relating to this species or higher taxon.
- **Universality:** The communication regarding scientific name. would be very difficult if we had only vernacular names in different languages; specialists would have to learn the name of taxa in innumerable languages in order to communicate with each other. To avoid this, zoologists have adopted by International agreement to use single language and a single set of names for each animals and its taxonomic groups. So, scientific names for taxa are universally accepted.
- **Stability:** If scientific names are frequently changes than the communication would be hampered. So, ICZN emphasized to retain the stability of scientific name by creating "Law of Priority".

Preamble: The objects of the Code are to promote stability and universality in the scientific names of animals and to ensure that the name of each taxon is unique and distinct. All its provisions and recommendations are subservient to those ends and none restricts the freedom of taxonomic thought or actions. Priority of publication is a basic principle of zoological nomenclature; however, under conditions prescribed in the Code its application may be modified to conserve a long-accepted name in its accustomed meaning. When stability of nomenclature is threatened in an individual case, the strict application of the Code may under specified conditions be suspended by the International Commission on Zoological Nomenclature. The International Commission on Zoological Nomenclature is the author of the Code.

Chapters of the ICZN: The "Preamble" is followed by 18 chapters which include 90 consecutively numbered "Articles".

Chapter 1: Zoological nomenclature (Article 1 to 3)

Chapter 2: The number of words in the scientific names of animals i.e bionomial nomenclature (Article 4 to 6)

Chapter 3: Criteria of publication (Article 7 to 9)

Chapter 4: Criteria of availability (Article 10 to 20)

Chapter 5: Date of publication (Article 21 to 22)

Chapter 6: Validity of names and nomenclatural acts (Article 23 to 24)

Chapter 7: Formation and treatment of names (Article 25 to 34)

Chapter 8: Family-group nominal taxa and their names (Article 35 to 41)

Chapter 9: Genus-group nominal taxa and their names (Article 42 to 44)

Chapter 10: Species-group nominal taxa and their names (Article 45 to 49)

Chapter 11: Authorship (Article 50 to 51)

Chapter 12: Homonymy (Article 52 to 60)

Chapter 13: The type concept in nomenclature (Article 61)

Chapter 14: Types in the family group (Article 62 to 65)

Chapter 15: Types in the genus group (Article 66 to 70)

Chapter 16: Types in the species group (Article 71 to 76)

Chapter 18: Regulations governing this code (Article 85 to 90)

Some Important Rules of Zoological Nomenclature

- Zoological nomenclature is the system of scientific names applied to taxonomic units (taxa; singular: taxon) of extant or extinct animals (Article 1.1).
- Zoological nomenclature is independent of other systems of nomenclature in that the name of an animal taxon is not to be rejected merely because it is identical with the name of a taxon that is not animal (Article 1.1.1).
- The scientific name above the sub-genera is uninominal (Art. 28).
- The names of species are binominal and those of sub species are trinomial.

- The scientific name above the sub genus is started wth capital letter and the species and sub species names are started with small letter.
- The first (Genus) and the third name (species) formed the binominal and the first, third and fourth (Sub species) formed the trinomial nomenclature.
- The presence of sub genus does not affect the status of nomenclature.
- The name of the genus, species and sub species are preferably italicized or underlined.
- The species are frequently are shifted from one genus to the other. Zoological Code has got provisions to govern such changes. Specific name can not be changed if it is valid, but the generic name can be changed.
- Comma is always used in between the author name and the year.
- **Law of Priority (Article 23) :** The valid name of a taxon is the oldest name applied to it. The date of publication of a name thus is of crucial importance. In zoological nomenclature the priciple of priority applies only to the categorial levels of species (and sub species) , genus, and family. It does not apply to the higher categories. The "Law of Priority" thus promote stability to the scientific name. It is restricted to the species (and sub species), genus and family.
- **Tautonyms (Article 18) :** The tautonym is a name of species or sub species in which the second or even the third components of the name repeats the generic name.
- **Power of the Commission :** Article 77 to 84 is devoted to the Power of the Commission.

Types of species level

- Species is unit of classification.
- It is necessary to understand different kinds of types recognized by the code for this category.

Holotype (article 73) - it is a single specimen designated by authors of the species as the name bearing type.

Syntypes – it is a series of specimen designated by the authors of the species as the name bearing type or indicated by the author (s) of the original description.

Lectotype (article 74) – it is one of syntypes designated as name bearing type of species by the first reviser.

Neotype - when holotype, syntypes and lectotype are lost or destroyed, then a specimen from type locality that agree with the original description is designated as the new bearing type.

Allotype: The opposite sex specimen which is described along with the holotype.

Paratype: Other specimens of the species kept along with the holotype and allotype.

Topotype: Specimen (s) collected from the same location as the holotype if the same name is given by different scientists to different organisms, it is called Homonymy

Types of speciation

- **Allopatric species:** The two or more related species that have separate geographical ranges are called allopatric species.
- **Sympatric species**: Two or more species are said to be sympatric when their geographical distributions overlap, though they may segregate into different ecological niche.
- **Parapatric species**: These are the species which have the geographical ranges with a very narrow region of overlap.
- **Peripatric species** when small group of individuals break off from the larger group and form a new species.
- **Sibling species:** Two or more than two closely related species which are morphologically alike but behaviourally or reproductively isolated from each other.
- **Cryptic species:** The species which are alike on the basis of observed features but are genetically and sexually they are different are cryptic species. There is confusion between the terms sibling species and cryptic species. The cryptic species are incapable of interbreeding.
- **Monotypic species:** When a genus includes a single species but does not include any subspecies, e.g., Vampyroteuthis, a vampire squid which is a single monotypic genus and also contains a single species, *V. infernalis* (monotypic species) .
- **Polytypic species:** When a species contains two or more subspecies, it is called polytypic species.
- **Endemic species:** The species which are found in a particular region, called endemic species. Usually the species of oceanic islands which are found in a limited geographic area are called endemic species.
- **Agamo species:** Species are those which consist of uniparental organisms. They may produce gametes but fertilization does not take place. They reproduce by obligatory parthenogenesis. In case of bees, wasps, rotifers the haploid eggs develop into haploid individuals and the haploid eggs are not fertilized by sperms.

Patterns of speciation

Fossil record shows two patterns:

Anagenesis (phyletic evolution): The transformation of an unbranched linage of organism, sometimes create an organism different enough to be a new species.

- In another words, anagenesis is an evolutionary change of a single lineage in which one taxon is replaced by another without branching.
- It is the evolution within a lineage.
- Anagenesis gives rise to paraphyletic taxa.

Cladogenesis: Branching evolution, budding of one or more new species from a parent species that continue to exist.

- Evolution with splitting of lineage.
- Cladogenesis gives rise to monophyletic taxa.
- The term cladoendesis was introduced by N. J. Kluge.

Cause of speciation

It often occurs when part of the population is isolated from another part.

PhyloCode

- The Linnaean system of binomial nomenclature is now becoming unsuitable to govern the naming of clade and species. Clade is a group in which every member shares a common ancestor (a unique common ancestor) . A clade is a group for which all the descendants of the last common ancestor of the members of the group are included in the group.
- Theoretical foundation of the phyloCode was developed by de Queiroz and Gauthier in a series of papers published in 1990, 92, 94.
- The first version of phyloCode was published in 2000.
- A group of cladists developed the Phylocode- a phylogenetic code of biological nomenclature, which is considered alternative to the Linnaean system.

Taxonomic publications

Species descriptions: We carryout initial identification depending on illustrations in field guide or handbook. Then the search of the unknown begins with thorough taxonomic descriptions. The style of writing taxonomic descriptions made and written varies from group to group. However, there is a common basic structure of Taxonomic descriptions. These descriptions include a heading that consists of scientific name, name, author, and date, followed by a synonymy, and then the main body of the description, which may include etymology, diagnosis, taxonomic discussion, ecology, and distribution sections.

Monograph: It is a comprehensive systematic study of a particular taxonomic group. Monographs try to review all the facts about a taxonomic group, species descriptions (includes morphology, phenology, ethnobotany, distribution, specimens cited, common names, etymology, etc.)

Checklist: A checklist provides a list of recognized taxa within a taxonomic group in a particular geographical area. A series of checklist provide details of all the taxa available at that particular time.

Catalogue: A catalogue can be defined as a complete list or a booklet or register, which comprises a list of characters and their alternates present in various taxa. The catalogue has special value in taxonomic studies and is also helpful in identifying species based on their features.

Websites related to insect taxonomy

antbase.org

Antbase is a collaborative effort between scientists from around the world, aiming at providing the best possible access to the wealth of information on ants, to fulfill the conservation needs of the International Union for the Study of Social Insects (IUSSI), and the Species Survival Commission of the World Conservation Union (IUCN). Antbase, together with the Hymenoptera On-line Database, is the data provider for ants to the Integrated Taxonomic Information System, ITIS. Antbase is being built and maintained at the American Museum of Natural History and the Ohio State University.

AntWeb

AntWeb are providing information and high quality color images of many of the approximately 10,000 known species of ants. AntWeb currently focuses on the species of the Nearctic and Malagasy biogeographic regions, and the ant genera of the world. Over time, the site will grow to describe every species of ant known. Taxonomic names in AntWeb are managed by AntCat.org Natural history information and field images are linked directly to taxonomic names. Distribution maps and field guides are generated automatically. All data in AntWeb are downloadable by users. AntWeb also provides specimen-level data, images, and natural history content to the Global Biodiversity Information Facility (GBIF) and Wikipedia.

Encyclopedia of Life (EOL)

EOL brings together trusted information from resources across the world such as museums, learned societies, expert scientists, and others into one massive database and a single, easy-to-use online portal. EOL collaborate with Biodiversity Heritage Library (BHL), Barcode of Life (BOLD),Catalogue of Life (COL), Global Biodiversity Information Facility (GBIF)

Integrated Taxonomic Information System (ITIS)

Taxonomic information on animals, plants, fungi, and microbes created in collaboration with several international agencies. A partner of Species 2000 and the Global Biodiversity Information Facility (GBIF). It provides authoritative taxonomic information on plants, animals, fungi, and microbes of North America and the world.

LepIndex

LepIndex is essentially a computerised and updated version of the Natural History Museum's (NHM), UK card index archive to the scientific names of the living and fossil butterflies and moths (Lepidoptera) of the world.

NCBI Taxonomy

(National Center for Biotechnology Information) The NCBI taxonomy database contains the names of all organisms that are represented in the genetic databases with at least one nucleotide or protein sequence. This database is searchable by both common and scientific names and provides lineages for many taxa as well as links to gene and protein data and other external Internet resources. This currently represents about 10% of the described species of life on the planet.

ScaleNet: A databse for the Scale Insect

The objective of this site is to provide comprehensive information on the scale insects (Coccoidea) of the world, including queryable information on their classification, nomenclatural history, distribution, hosts, and literature.

ZooBank

ZooBank is intended as the official registry of Zoological Nomenclature, according to the International Commission on Zoological Nomenclature (ICZN).

Zoological record

In 1864, a group of scientists affiliated with the Zoological Society of London and the British Museum founded Zoological Record as a way to communicate amongst them. As others in the scientific community started to use this resource, its content was expanded and the Zoological Society of London assumed complete responsibility for its publication as of 1886. In 1980, the Zoological Society and BioSciences Information Service of Biological Abstracts™ (BIOSIS™) joined forces to produce and publish Zoological Record, and today, BIOSIS (part of the Web of Science) is the sole publisher.

MOLECULAR SYSTEMATIC

Chemotaxonomy

- Candolle (1813) was first to approach this method in his experiment and to develop new methods to identify closely related species.
- It was well known that the metabolic activities of an organism are related with complex chemical changes.

Cytotaxonomy

Cytotaxonomy deals with all the aspects of taxonomy at a cellular level and its includes, the structural, genetical and biochemical aspects and is used to compare genetic affinity between organisms. This helps in the identification of organisms.The type of cytological data commonly used in systematics are basically from.

- DNA hybridisation studies are used to study and compare the genetic affinity between two organisms. DNA of both organisms is extracted in this tqchnique and estimating the genetic similarities among them.
- Karyological studies: here chromosome number and structure which are used quite extensively either separately or both together.

Molecular taxonomy: Molecular taxonomy is the classification of organismson the basis of the distribution and composition of chemical substances in them.The name molecular taxonomy was given by Lahni (1964) and Turner (1966) preferred to divide into two types.

- Micromolecular taxonomy
- Macromolecular taxonomy

DNA Barcoding

- The concept of molecular barcoding was proposed by Floyd *et al.* in 2002.
- It is developed into a global biological identification system by Paul Hebert *et al.* in 2003.
- DNA barcoding rapidly and accurately identify a species by amplifying and sequencing a short, standardised region of its genome, specifically the mitochondrial gene, cytochrome c oxidase subunit 1 (COI) by Paul Hebert *et al.*, 2003.
- This method depends on DNA and not morphology; it is applicable to any life stage, from egg to adult.
- It also aims to employ standardised protocols that may be applied to a wide range of organism.

Drawbacks / weakness of DNA barcoding

- No universal DNA barcode gene
- Single gene approach
- Difficult to resolve recently diverged species
- Identifies inter specific genetic variation only

History of insect classification

- The evolution of modern classification of insects is often studied under four subheadings as given by Engel and Kristensen in 2013 and included 4 stages i.e the Pre-Linnean Era, the Linnean Era, the Darwinian Era, and the Hennigian Era.
- Aristotle (384-322 BC) grouped all flying insects with other flying animals like bats and birds which reflects huge artificiality in his classification.
- Aldrovandi in 1602 distinguished insects based on their habitat mainly land and water and he attempted classification of insects based on wing and leg morphology.
- Thomas Mouffet (1553-1604) attempted to classify insects based on their habits in his book Insectorum sive Minimorum Animalium Thetrum.
- Classification of insects based on their morphology, biology, ecology and anatomy was produced by John Ray (1627-1705) in his book Historia Insectorum.
- Linnaeus (1707-1778) put forth a systematic classification of insects in 10 edition of his book wherein the bionomial nomenclature as the convention of naming the organisms was popularized and also classified insects on the basis of presence or absence and the number of wings present in adult insect.
- Johan Christian Fabricius (1745-1808) classified insects on the basis of mouth parts and he regarded it as a more natural character than number of wings and published his system of classification in his books Systema Entomologiae and Genera Insectorum
- Pierre and Relatreille (1762-1833) proposed the classification of insects that is considered first truly natural classification of insects
- Ernst Heinrich Philipp August Haeckel (1834-1919) for the first time proposed an explicit phylogeny of the insect orders in his book Generelle Morphologie
- Friedrich, M. Brauer (1832-1904) was the first to classify insects based on Darwinism He recognised 16 orders

- Classification of insects divided winged insects into Exopterygota and Endopterygota on the basis of external or internal development of wings that was adopted by August D. Imm (1881-1949) too in his influential book Gerenal Text Book of Entomology into 29 orders.
- Modern classification of class Insecta up to the rank of orders has been taken from BARNARD in 2011 which as 28 Orders. Here, order Dictyoptera includes three suborders, Blattodea, Mantodea and Isoptera.
- Another recent classification of insects given by Gary Parsons in 2015 and included 27 orders under class Insecta.
- Willi Hennig gave new term to order Thysanura into Zygentoma. Whereas order, Grylloblattodea and Mantophasmatodea included under the order Notoptera. And, order Isoptera included into Blattodea.

1.2 Arthropoda

Character of phylum arthropoda

- The body segmented into two or three distinct region
- Jointed appendages
- Bilateral symmetry
- Triploblastic animals
- Chitinous exoskeleton, which is shed and renewed periodically as the animal grows
- A tubular alimentary canal which extend from mouth to anus
- An open circulatory system, no blood vessels, the organs directly bathed in blood
- Heart dorsal in position
- Excretion by means of green gland and coxal gland in aquatic and malphigian tubules in terrestrial insect
- The body cavity a blood cavity or hemocoel, the coelom reduce
- The nervous system is ventrally located and consist of an anterior ganglion and posterior ganglion
- Respiration by means of gills or tracheae and spiracles
- No cilia or nephridia
- The sex always separate
- Development either direct or indirect
- Cleavage should be meroblastic and superficial
- Types of eggs, centrolecithal

Classification of Arthropoda

Sub phylum: Trilobita (Extinct arthropods)

- They lived during the Paleozoic periods (but abundant during Cambrion and Ordovician periods)
- Elongated and flattened animals with three distinct longitudinal division of the body.
- They had a pair of antennae
- Biramous appendages
- The anterior part of body cover with a carapace

Sub phylum: Chelicerata

- Absence of antennae
- Typically have six pair of appendages, 1st pair are the chelicerae and rest are leglike
- The body divided into two distinct division i.e prosoma (cephalothorax) and opisthosoma (abdomen)

Sub phylum: Mandibulata

- Mouth part mandibulate type
- 1st pair of appendages called Antennae
- Uniramous or biramous appendages appendages

1.3 Hexapoda

Characteristic features

- Body three distinct region i.e head, thorax and abdomen
- One pair antennae (rarely absent)
- One pair of mandible and maxillae
- A hypopharynx
- A labium
- Three pairs of legs of each thoracic segments (some insect legless and some larvae have legs like appendages called prologs)
- The gonophores (rarely two) on the posterior portion of the abdomen
- No locomotors appendages on the abdomen of the adult (except in some primitive hexapods)
- At one time, all six-legged arthropods were considered "insects". But as we continue to learn more about these fascinating creatures, it has become increasingly difficult to justify grouping all "six-leggers" in a single class.

- There are major differences in external structures, organ systems, and post-embryonic development that seemingly point to evolutionary divergence between the "true" insects and three other closely allied groups: Protura, Diplura, and Collembola.
- Many entomologists still consider these three taxa to be "primitive" orders of insects, others prefer to classify them as orders in a separate class (the Entognatha) , and still others insist that differences outnumber similarities and regard each taxon as a separate class of arthropods.
- This is the perennial debate between "splitters" and "lumpers". The argument is not new and it will not be settled quickly. In this writing, it is chosen to recognize Elliplura, Diplura, and Insecta as three separate classes of hexapoda.

Super class Hexapoda (six legs) divided into 3 classes

Class 1. Ellipura

- Order: Collembola (Springtails)
- Order: Potura (Proturan

Class 2. Diplura (Dipurans)

Class 3. Insecta

Classificationsof class Insecta

Classes: Insecta

Sub class

1) Apterygota

Orders

- Archaeognatha (Jumping bristletails)
- Thysanura (Silverfish)

The classification is shown through diagram

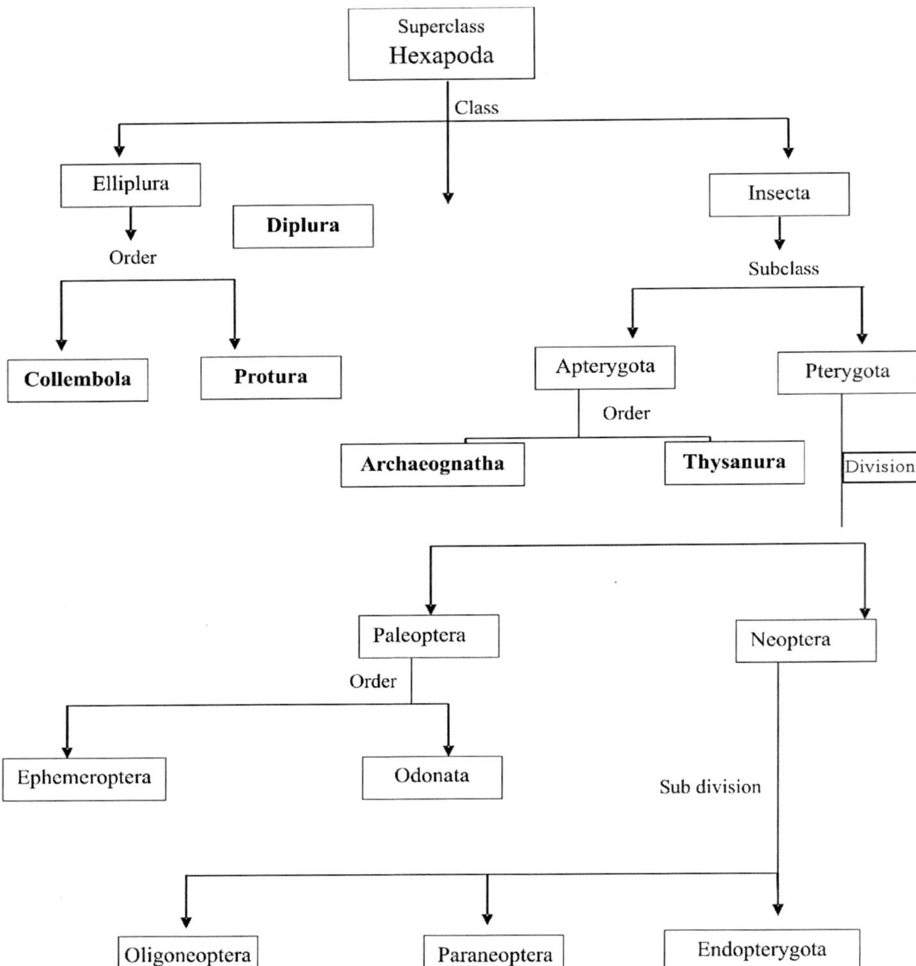

Classification of super class hexapoda on the basic of phylogeny

2) Pterygota

Division

1. Paleoptera (They cannot flex their wing over the abdomen)

- Order: Ephemeroptera (May flies)
- Order: Odonata (Dragonflies, Damselflies)

2. Neoptera (New wings) (These insect can flex their wing over the abdomen.)
   Sub division

a. Orthopteroid and blattoids orders (oligoneoptera): They have diverse type of mouth part and varieties forms of wings.they have large no of veins in the wings.

Orders

- Plecoptera (Stoneflies)
- Blattodea (Cockroach)
- Isoptera (Termites)
- Mantodea (Praying mantids)
- Grylloblattodea (Rock crawlers, Ice bug)
- Dermaptera (Earwigs)
- Orthoptera (Grasshopper, Locust)
- Phasmatodea (Stick and Leaf insect)
- Mantophasmatodea (Gladiators)
- Embioptera (Webspiners)
- Zoraptera (Angel insect)

b. Hemipteroid orders (Paraneoptera)

- Psocoptera (Book and Bark lice)
- Pthiraptera (Chewing and Sucking lice)
- Thysanoptera (Thrips)
- Hemiptera (True bugs, Hoppers, Scales)

c. Endopterygota

Section

i) Neuropteroid –Coleopteroid orders

- Strepsiptera (Twisted wing parasite)
- Megaloptera (Dobsonflies, Alderflies)
- Raphidioptera (Snakeflies)
- Neuroptera (Antilions, Lacewings)
- Coleoptera (Beetles)

ii) Panorpoid orders

- Mecoptera (Scropianflies)
- Siphonaptera (Fleas)
- Diptera (True flies)
- Trichoptera (Caddisflies)
- Lepidoptera (Moths and Butterflies)

iii) Hymenopteroid orders

- Hymenoptera (Bees)

2

Entognatha

Characters

- Three orders i.e. protura, diplura and collembolan are group together as Entognatha because the mouth parts are withdrawn into the head
- Antennal flagellum (if present) are masculated
- Tarsi one segmented
- Compound eye either absent or ommatidia reduced in number
- Tentorium rudimentary

2.1 Protura (Pro-first, ura-tail)

It refers to the lack of advanced or specialized structures at the back of the abdomen. These are most primitive wingless of all hexapods and lack metamorphosis

Ex. Proturans (nick named cone head)

Habitat: They live in moist soil or humus, in leaf mold, under bark and in decomposing logs. They feed on decomposing organic matter and fungal spores and found worldwide.

Characteristic feature

- They are minutes whitish hexapods, 0.6 to 1.5 mm in length
- Head is somewhat conical and bearing two psuedoculi which are sensitive to light
- Absent-eye wings and antennae
- Mouth parts piercing and sucking type
- 1st pair of legs act as sensory function (serve as antennae)
- Tarsi one segment
- Cerci absent
- Trachea usually absent
- Malphighian tubules papillae like

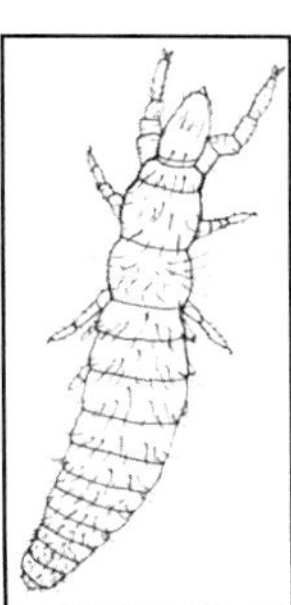

- Styli are present on 1-3 abdominal segment.
- On hatching from eggs, the abdominal segment consist of 9 segment and after each next three moult segments are added anterior to the apical portion (telson). This process is known as anamorphosis.
- Adult have 12 abdominal segment (11 metameric segment and the one telson).
- Body unpigmented, usually white or ivory in color.

2.2. Diplura (Di-two, plura-tails)

It refers to the large cerci at the rear of the abdomen

Two prolonged bristle tails

Ex. Japyx

Habitat: They are found in damp situation in caves, under tree bark, in the soil and it is cosmopolitan.

Characteristic features

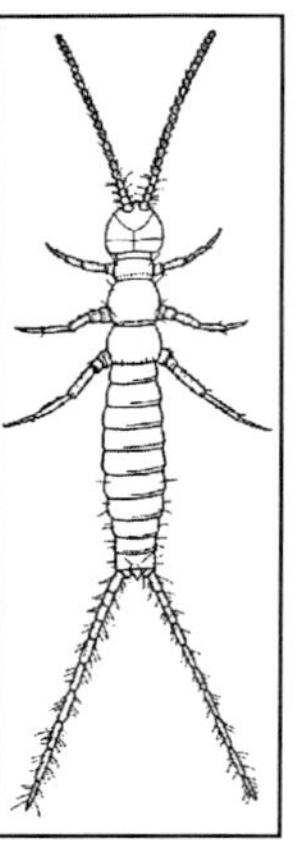

- They are 5 mm to 50 mm in length
- They are similar to silverfish and bristletails except lack of median caudal filament
- Body is not covered with scale
- Narrow, elongated body usually colourless
- Mouth part biting and chewing type
- Antennae long many segmented moniliform type
- Compound eye and ocelli are absent, but may have light sensitive spots on heads called pseudoculi.
- Tarsi one segment
- Styli are present on 1-7 and 2-7 abdominal segment.
- They have small, eversible vesicles on the ventral side of most abdominal segments that help to regulate the body's water balance, perhaps by absorbing moisture from the environment
- Abdomen 11 segment
- Last abdominal segment posses a pair of cerci, In one common family (Japygidae which is pincer like in carnivorous proturans.

2.3 Collembola (Colle-glue, emblos-piston or peg)

Ex. Springtail

The name Collembola, derived from the Greek "coll" meaning glue and "embol" meaning a wedge, refers to a peg-shaped structure, the collophore, on the underside of the first abdominal segment. The collophore function as an adhesive organ.

Habitat: They are found in soil, leaf litter and decaying vegetables matters and soil fungi. Most are saprophagous in nature

Characteristic features

- Collembolan is derive from colle means glue and embolon means piston or peg.it refers to the belief that the ventral tube or collophore has adhesive properties and glue-peg function as maintain water balance and excretion.
- It is about 6 mm long.
- Body frequently clothed with scales
- Biting mouth part
- Antennae 4 sement
- Abdomen has 6 segment

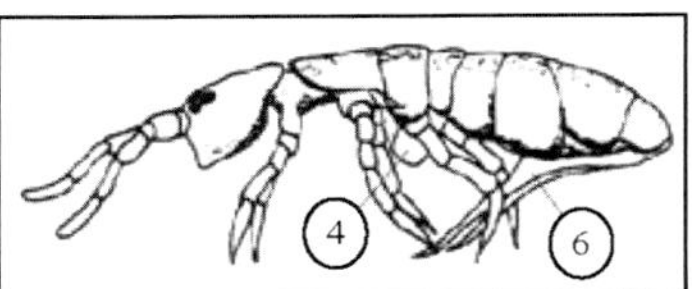

- 1st abdominal segment known as collophore or ventral tube (4)
- 3rd abdominal segment bear a pair of retinaculum or hamula
- Releasing the tenaculum causes the furcula to snap down against the substrate and flip the organism some distance through the air.
- This device, present in all but a few genera, seems to be an effective adaptation for avoiding predation.
- 4th abdominal segment bear a jumping organ or furca (6). It is forked like structure present on ventral side.
- Both abdominal segment/appendages together form springing apparatus/ mechanism
- Lack of tracheal respiratory system and exchanges of gases though cuticle or integument.
- External genetalia ,cerci absent
- Immature collembola are similar in appearance to adults.
- They usually moult 4-5 times

Economic Importance

Springtails are part of the community of decomposers that break down and recycle organic wastes. A few species feed on living plants and are occasionally regarded as pests: *Bourletiella hortensis* (the garden springtail) may damage seedlings in early spring, *Sminthurus viridis* (the lucerne flea) is a pest of alfalfa in Australia, and *Hypogastrura armata* has been a frequent pest of commercial mushrooms.

3

Ectoganatha (Apterygota)

Characters

- Two orders i.e. Archaeognatha & Thysanura are grouped together as Ectognatha because the mouth parts are more or less exposed and not covered by cranial fold.
- Species of two order are small to medium sized creatures with simple metamorphosis and it is closest living relatives of the winged insect
- Antennal flagellum are not masculated
- Tarsi 3-5 segmented
- Compound eye usually present
- Tentorium fairly well develop
- Abdomen has three long caudal filament

3.1 Archaeognatha (Archaeos- ancient/old,gnatha-jaw)

Ex. - Jumping bristletails

Name refer to articulation of mandible into head capsule in single point and this order also known as Microcoryphia.

Habitat: They are nocturnal and found in wide variety of habitat. They prefer damp environment and may be found under the bark of tree, in soil or leaf litter or in the rocks cervices.

Characteristic features

- Bristletails are mostly grey in colour and usually less than 20 mm in length.
- They are more cylindrical with thorax somewhat arched
- Elongated body that tapers towards the back of abdomen.
- They can able to jump up to 10 cm by snapping abdomen against ground.
- Body covered with minute scale
- Compound eyes are large and contiguous
- Ocelli always present
- Long antennae with many small segment
- 3 long slender cerci with the middle one much longer than two out side

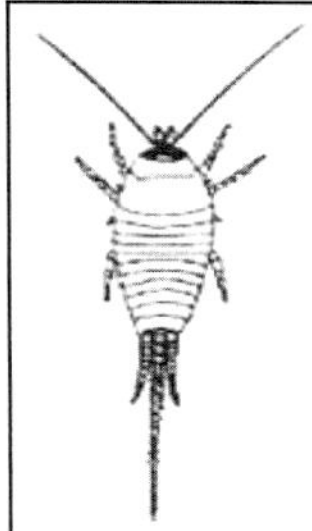

- Wingless
- The abdomen bear a pair of styli on segment 2-9 and 2-7
- Segment 1-7 usually bear 1 or 2 pair of reversible vesicles on the abdomen function as water absorbing organ.
- Sexual maturity is reached after at least eight juvenile instars spanning up to two years.
- Molting continues periodically even after adulthood.
- The sexes are separate, but copulation does not occur.
- Males produce a packet of sperm (spermatophore) and leave it on the ground to be picked up by a female.
- Females cannot store sperm (they lack a spermatheca), and evidently acquire a new spermatophore before each bout of egg laying.
- Eggs are laid singly or in small groups (<30).

3.2 Thysanura (Thysan-tassel/bristle/fringed, ura-tail)

Thysanura is also known as Zygentoma

Ex. Silverfish, Firebrat

Habitat: They are found in household items, museum and library it is found in cool, damp location where as firebrats prefer warmer location. They feed upon book binding, wall paper and decaying matters.

Characteristic features

- They are very fast running insect
- Elongated and somewhat flatten with three tails like appendages at posterior of abdomen
- Body always covered with scale
- Mouth parts mandibulate type adopted for biting and each mandible has two point of articulation with head capsule
- Flightless insect
- They have long filiform antennae with more than 30 segment
- Tarsi 3-5 segment
- Tracheal system well developed.
- Compound eyes are small and widely separated (or absent)
- Abdomen 11 segmented and last abdominal segment hve 2 cerci and 1 filament like structure called telson.
- Styli form appendages located on abdominal segment 7-9.

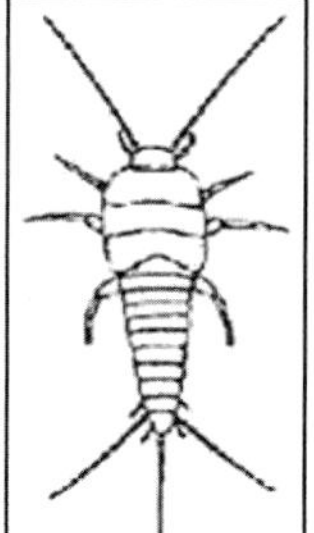

Important order

Lepismatidae: *Lepisma saccharina* (silverfish)

Themobia domestica (firebrat)

4

Palaeoptera

Characters

- Palaeoptera world derive from New latin word pale means old and ptera means wings
- Two order i.e Ephimeroptera and Odonata comes under this grops because wings can not fold over abdomen at time of rest.

4.1 Ephemeroptera (Ephemero-a day or short lived, Ptera - wing)

Ex. May flies

Habitat: Nymph stages are aquatic and adults are terrestrial or aerial. May flies are an important food of many fishes, bird, amphibian, spider and other animals. May fly funna in aquatic habitat may serve as an indicator of ecological characteristic of that habitat.they generally feed on algae but some are predatory.

Characteristic feature

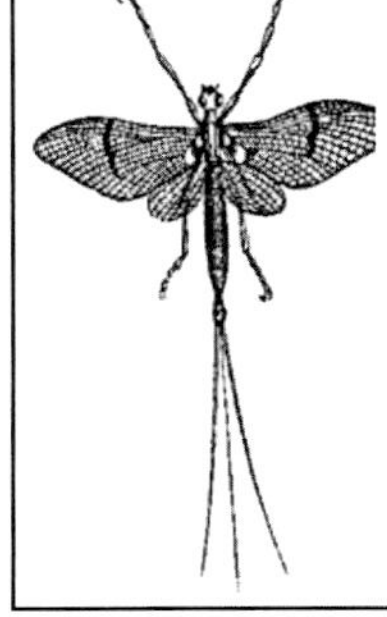

- It is small to medium sized,elogate,very soft bodied insect with 2 or 3 long thread like tails.
- They are aquatic insect because they lay eggs, grow and develop in water.
- Nymph known as Naids and subimago is last stage of nymph after one moult become adult.
- Subimago is covered with very fine hairs called pellicle.
- Adult have membrainous wings with numerous veins. The front wing usually large and triangular and hind wings are small and rounded.
- The life span of adult insect is very short and survive 30 minute to one day.
- They only perform the function of reproduction.
- The adult of may fly the mouthparts are vestigial, while nymph chewing type
- They have compound eye and 3 ocelli.
- Short setaceous type antennae

- Tarsi 3-5 segmented, bilateral respiratory gills on 1st 4-7 abdominal segment.
- Their alimentary canal is filled with air
- They usully recognized by the leaf like or plumose gills along the sides of abdomen
- They moult 20-30 times

Importance: May flies are unique among the insect that is passing through a winged Sub adult/subimago stage. It is identical to the adult stage except lacks of functional genitalia. The insect forms important link in food chain of aquatic ecosystem.

4.2 Odonata (Odontos-Tooth)

- Odonata name derive from greek word,meaning tooth i.e mandible have strong tooth.

Ex. Draginflies and Damselflies

Habitat: All stages are predaceous and feed various insect and organism.

Characteristic features

- They are large and beautiful coloured insect with two pair of membranous, many veined wings and abundant in the Oriental regions.
- At the anterior portion of the wing area which os known as pterostigma. This is thickened blood filled on after colourful area and bounded by veins.
- Dragonfly and damselfly are peculiar amongst insect in having copulatory organ of male located at anterior end of abdomen on the ventral side of II and III segment. Before copulation, males transfer the sperm from the genital opening to the functional penis. This transfer is achieved by bending of abdomen. The male genetalia of other insect are located at the poseterior end of abdomen.

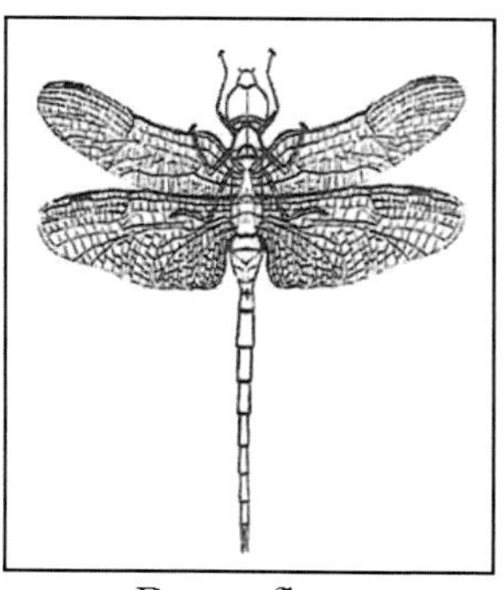

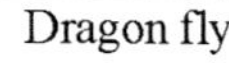

Dragon fly

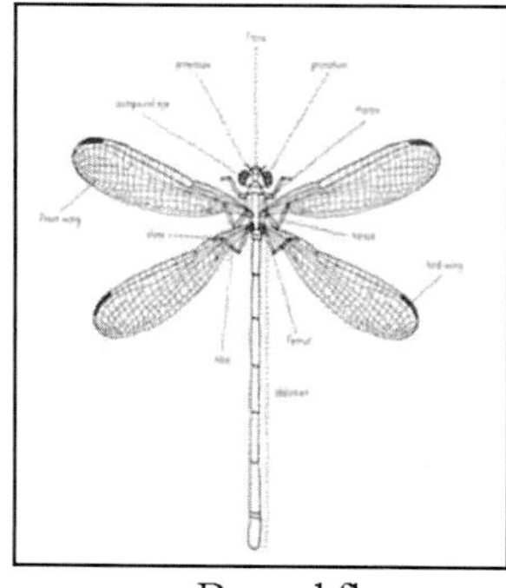

Damsel fly

- The copulatory organ of male odonata are therefore considered as secandary genetalia.
- The immatures stages are aquatic in nature and adults are found near water
- The compound eyes are very large and three ocelli
- The antennae are very small / filiform and bristle like.
- Mouth parts are chewing type.
- Cerci unsegmented and function as claspers as female.
- Thorax consist of a small prothorax and large pterothorax.
- In case of nymph, the labium is modification of prehensile organ for grasping prey.
- Respiration by means of rectal or caudul gill.

Class odonata further divided into two sub-order as follows:

Sub order : Anisoptera	**Sub order : Zygoptera**
1. Fore wing and hind wing are unequal size and hind wing basally broader than fore wing.	Fore wing and hind wings are same size and shape with wing bases narrow gradually.
2. The wings are held horizontal at rest.	The wings at rest held vertically
3. The males have the appendages at the apex of the abdomen. Two cerci and epiproct are present	The male have four appendages at the apex of the abdomen. Two cerci and a pair of paraprocts.
4. It is strong flier	It is weak flier
5. Compound eyes large and holoptic.	Compound eyes button like and dichoptic.
6. Male has three abdominal appendages (paraproct absent).	Male has four abdominal appendage (epiproct absent).
7. Oviposition exophytic	Oviposition endophytic
Naiads	
8. Nymphs have five short stiff appendages at the apex of abdomen and have epithelial gills in the rectum.	Nymps have three gills at at the apex of abdomen
9. Stout & robust.	Slender & fragile.
10. Rectal gills (internal) found.	Caudal gills (external) found.
11. Jet propulsion mechanism present	Jet propulsion mechanism absent

5

Oligoneoptera

Characters

- Mouthparts biting type
- Wing venation usually well developed with numerous cross veins.
- Cerci present
- Terminalia of male may be asymmetrical and reduced
- Many Malpighian tubules

5.1 Plecoptera (Pleco – folded/pleated/twisted, ptera wing)

Ex. Stonefly

Habitat: They feed on decaying vegetation and they are found under stones in streams or along lake shores. Nymphs as pollution indicators because development depend upon well oxygenated water.

Characteristic features

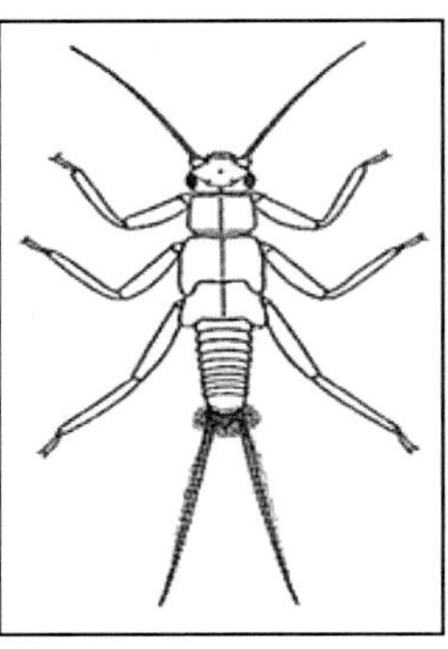

- They are mostly medium size or small somewhat flattened and soft body insect
- Most species have four membranous wings and hind wings are slightly shorter than the front wings but some species wingless or they have short wing known as brachypterous.
- Adults rarely feed, possibly on algae, pollen and lichen, nymphs are phytophagous or carnivorous
- 4 – 50 mm long.
- They have well developed anal lobe that is folded fanwise when the wings are at rest.
- Antennae setaceous type long, slender and many segmented
- Tarsi 3 segmented
- Cerci (multiarticulated) are present and may be long or short.
- Mouth parts are of chewing type (reduce in adult and do not feed)
- They are hemimetabolous exopterygota insects

- Nymph are aquatic with branched gills on the thorax
- Adults are terrestrial
- Lay eggs in water and nymph may moult 10-30 times.
- Ovipositor wanting
- Lack median caudal filament

Difference between stonefly and mayfly

Stonefly	Mayfly
1. They are very similar to mayfly of nymph but lack a median caudal filament.	Present
2. They have only two tails	Always have three
3. nymph of stone fly have two tarsal claws	mayfly nymph have only one
4. Stonefly nymph have fingerlike gills and occur ventrally	Mayfly nymph have leaf like gills along the side of the abdomen

5.2. Blattodea (Blatta light avoiding insect)

Ex. American Cockroaches- ***Periplaneta americana***

German cockroach – ***Blattella germanica***

Orientalis cockroach –*Blatta orientalis*

Habitats: They are common in tropical and sub-tropical climate and they are omnivorous and nocturnal in nature. They feed on all sorts of things in a house and contaminate food. Desert to semi-aquatic, mainly in humid places, including human dwellings, caves and ants' nests. Scavenge mostly organic matter.

Characteristic features

- Earliear Mantodea, Isopteran and Blattaria comes under order Dictyoptera
- They have oval broad dorso-ventrally flattened body and relatively small head and covered by pronotum.
- Cockroaches are cursorial insects and they run very fast.
- Tarsi with 5-segment
- Cerci one to many segmented
- Antennae are long and filiform
- Eggs usually contained in ootheca or capsule.
- Leathery fore wing called tegmina
- Male genitalia asymmetrical
- Two ocelli (fenestrae)
- Wing buds of the Nymph do not undergo reversal

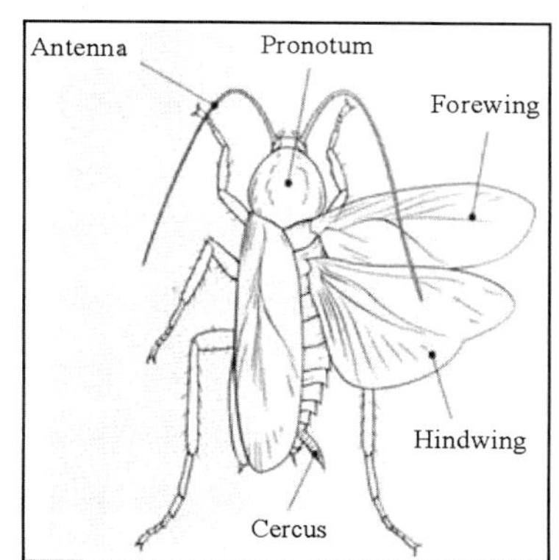

5.3 Isoptera (Iso- equal, ptera –wing)

Ex. Termites / White ants

Habitats: Termites lives in moist subterranean habitats and other live in dry habitats aboveground and subterranean forms normally live in wood buried or in contact with soil

Characteristic features

- Soft bodied insect.
- Winged member with two pairs of long narrow wing with similar form size and texture.
- These are social insect of which different cast lives together in colonies.
- Mouth parts chewing type.
- Head prognathous.
- Antennae short and monilliform type.

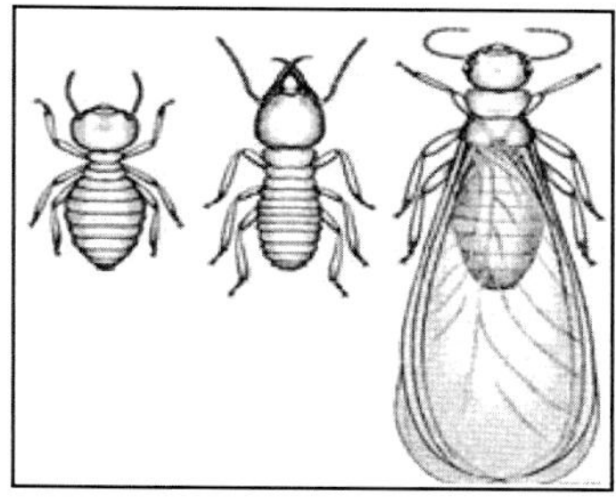

- Tarsi 4 segment
- Compound eye and ocelli are absent (if present two in number)
- Cerci is small and sometime rudimentary
- Frontal gland is a characteristic termite organ which attains its greatest development in soldiers.
- It is formed by a group of hypodermal cells in the median line of frons.
- It is sac like gland which communicates to the exterior by frontal pore, which opens in a shallow depression, on the surface of the head where the cuticle is pale, which is known as frontanella. It appears to have defensive functions
- Large number of intestinal symbionts is housed in rectal pouch.
- Members show oral trophallaxis (mouth to mouth) and anal trophallaxis (anus to mouth) transfer of food.
- Queen shows physogastry- enlargement in abdomen due to egg development.
- Cellulose digestion- For cellulose digestion lower termites is dependent on flagellate protozoans and higher termites depend on fungus and bacteria or produce digesting enzymes themself.
- The family Macrotermitinae (Termitidae) develops fungal gardens called Termitomycetes where workers cultivate the fungus on a special substratum derived from its excreta. The fungal fruiting bodies are important source of vitamins and organic nitrogen.

Termites' castes

- Reproductive form, alate king and queen develop a colony.
- The worker and soldier castes made up of both sexes are sterile wingless.
- The worker are most numerous individual in a colony, they perform most of the work of the colony like nest building and repair for the colony.
- The soldier has enlarged head and stout mandibles adopted for defense.

Difference between reproductive and non-reproductive member

Reproductives	Non-reproductives
Late stage nymph with developing wing buds differentiate into reproductives **Primary:**Body dark, well sclerotised, well developed compound eyes and wings, original founder of colony **Secondary:** Poor sclerotization, pale body, compound eyes and wings not well developed. Replace primary reproductives	Larvae without developing wing buds differentiate into non reproductives.Sterile, apterous

Difference between caste

Queen	King	Worker	Soldier
Single queen. Queen shows physogastry.	Fertilizes queen. Helps in construction of nuptial chamber	Well-developed mandibles and salivary gland. Feed the king, queen, soldiers; care for eggs and young ones. Build and repair termataria, go out for foraging, cultivate fungus garden	Defend the colony. Mandibulate soldiers- large well sclerotised head with mandibles. Nasute soldiers- head drawn into nozzle shaped projection.

Families

1. Mastotermitidae:

- tarsi 5-segmented
- hind wing well developed with large anal lobe
- thorax broad and a pair of ocelli
- true worker are absent

Ex. *Mastotermes drwiniinsis* (subterranean termite)

2. **Kalotermitidae:**
- Commonly called dry and damp wood termites.
- True worker cast is absent
- Ocelli are always present
- Pronotum flat usually broader than head.
- Tarsi 4-segment
- Anterior wing scale long

  Ex. *Cryptotermes domesticus*

3. **Termitidae:**
- Commonly called ground dwelling termites
- Worker caste well developed
- Pronotum of worker and solider narrow.
- Compound eye present
- Wings very slightly reticulated and anterior wing scale short.

  Ex. *Odeontotermes obesus*

  Microtermes sp.

4. **Hodotermitidae:**
- Commonly called rotten wood termites
- Worker are present in some genera
- Ocelli are absent
- Pronotum saddle shaped
- Anterior wing scale short

  Ex. *Hodotermes* sp.

5. **Rhinotermitidae:**
- Commonly called subterranean termite
- Worker castes are present
- Pronotum of worker and solider is flat
- Anterior wing scale long
- Compound eye present

  Ex. *Rhinotermes* sp.

5.4 Mantodea (Mantodea mantis)

Ex. Mantids

Habitats: These insects are highly predaceous and feed on variety of insect and act as biological control agent. They usually lie in wait for their prey with the front legs in an upraised position.

Characteristic features

- Mantids are large, elongated and slow moving insect that are striking in appearance because of their peculiarly modified front legs.
- Fore legs raptorial type adopted for preadation having elongated coxae and spined femur and tibiae.
- Fore wing heavier in texture than hind wings
- Wing pads not reversing.
- Entire prothorax elongated and meso and matathoracic segments short
- Specilized auditory and stridulatory organ absent
- Small triangular head and deflexed, movable with neck
- Biting mouth part and filiform antennae
- Abdomen flattended, Cerci short and segmented.
- The mantids exhibit cryptic colouration thus simulating well with the background.
- The eggs are laid in water tight egg cases which are fixed to the plants.
- Pronymphal stage present. The case is prepared from a frothy gum, secreted by the female.
- After exposure to air, the case hardens and nymphs resembling ants emerge from them.

  Eg: *Gongylus* sp.

  Mantis religiosa

5.5 Grylloblattodea (Gryll-cricket and blatta - cockroach)

Ex. Ice bug or Ice crawlers

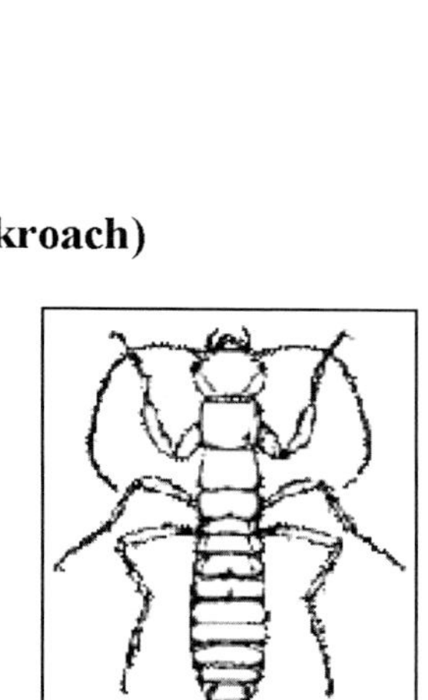

Habitats: They live beneath store or caves of high mountains and mostly nocturnal preferring low temperature below 1^{0}C. The black eggs are deposits singly in the soil when female is about one year old. Eight nymphal instars take five years.

Characteristic features

- It refers to the blend of cricket-like and roach-like traits found in these insects.
- Apterous orthopteroid insect without eyes and ocelli.

- Lacking in india
- Head prognathous
- Antennae long filiform
- Mandibulate type mouth part
- Prothorax longer than meso and metathorax.
- Wing absent and leg cursorial type
- Tarsi 5-segment
- Ovipositor well developed and sword shaped.
- Cerci long and 8-segment

5.6 Dermaptera (Derma-skin, ptera-wings)

Ex. Earwigs

Habitats: They are ground dwelling, in crevices, scavenge plant and animal matter. Nocturnal and omnivorous in habit. It is predatory and parasites on other insect. It also infests plant into developing bud and fruits.

Characteristic features

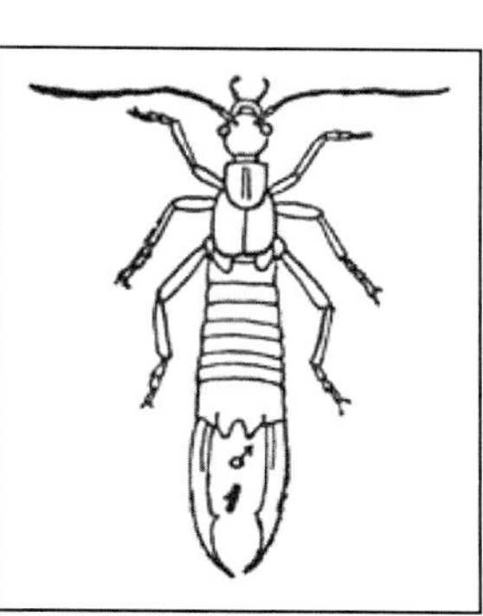

- Small to moderate sized rather flattened insect
- Head prognathous
- Antennae setaceous or monoliform.
- Compound eye well developed but ocelli absent.
- Fore wings reduced to small leathery flap (tegmina or elytra) .
- Hind wing membranous and semi-circular folded under forewings at rest
- Many species wingless also.
- Cerci unjointed and modified into prominent Horney forceps.
- Mouth part biting type.
- Development hemimetabolous
- Earwigs overwinter in the adult stage.

5.7. Orthoptera (Ortho - straight; Ptera-wings)

Ex. Grasshoppers, Locust, Katydid, Cricket, Mole cricket

Habitats: Desert to caves, in trees and subterranean burrows.They are scavengers, omnivores, herbivores and carnivores.

Characteristic features

- They are medium to large sized insects.

- Antenna is filiform.
- Mouthparts are mandibulate.
- Prothorax is large; pronotum is curved, ventrally covering the pleural region.
- Hind legs modified for jumping and saltatorial type.
- Forewings are leathery, thickened and known as tegmina. They are capable of bending without breaking.
- Hind wings are membranous with large anal area. They are folded by longitudinal pleats between veins and kept beneath the tegmina.
- Wings pad undergo reversals. Cerci are short and unsegmented.
- Ovipositor is well developed in female.
- Metamorphosis is gradual. In many Orthopterans, the newly hatched first instar nymphs are covered by loose cuticle and are called pronymphs.
- Specialized stridulatory (sound-producing) and auditory (hearing) organs are present.

Classification

This order is sub divided into two suborders, viz., Caelifera and Ensifera.

S.No.	Caelifera	Ensifera
1.	Antennae shorter than body.	Antennae long as body length and many segmented
2.	Auditory (tympanal) organ present on the base of abdomen.	Audatory (tympanal) organ present on fore tibiae.
3.	Stridulatory organ present femoro alary region.	Stridulatory organ present usually tegminal.
4.	Tarsi 3 sgmented	Tarsi 3 or 4 segmented
5.	Ovipositor short and robust	Ovipositor elongated, needle like or sword shaped.

Sub order: Caelifera

Family: Acrididae

Ex. Locusts, Grasshoppers

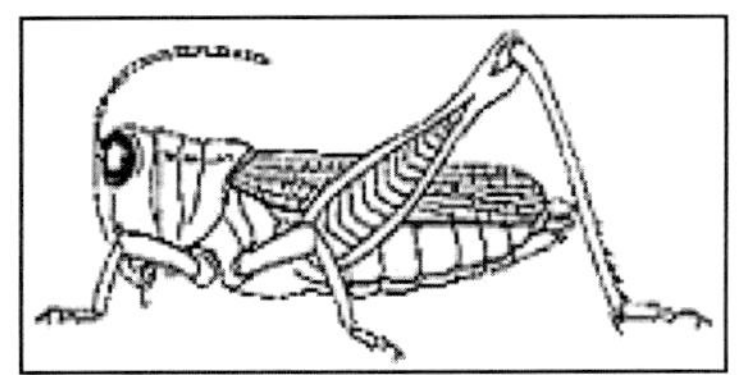

- Antenna is short
- Tarsi is 3 segmented
- Ovipositor is short and horny
- Tympanum is located one on either side of the first abdominal segment.
- Thorax covered with saddle shaped pronotum.
- Sound is produced by femoro-alary mechanism.

- A row of peg like projections found on the inner side of each hind femur is rubbed against the hard radial vein of the closed tegmen.
- Locusts are a serious threat to tropical agriculture. They swarm under favourable conditions and mainly feed on grasses, cereals etc.

Phases of Locusts

Gregarious: It is characterized by development of gregariousness and migration. The colour of nymph is black, yellow and orange and adult have shorter, saddle shaped pronotum with shorter femora.

Solitary: It is characterized by isolation, sedentary and no migration. The color of nymph is similar to normal enviroment i.e green, gray and brown with hind femora larger as compare to fore wing.

Sub order: Ensifera

Family: Tettigonidae

Ex. Katydids, Long horned grasshppers

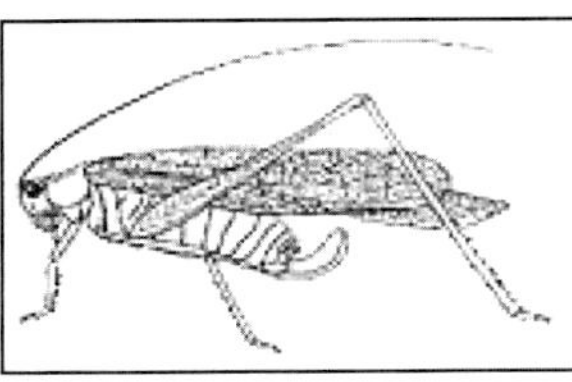

- Antenna is thread like as longer than body.
- Females with blade-like ovipositors
- They are Omnivorous and carnivorous
- They are also Crepuscular and nocturnal
- Stridulate using wing bases.
- A thick region on the hind margin of the forewing (scraper) is rubbed against a row of teeth on the stridulatory vein (file) present on the ventral side of another fore wing which throws the resonant area on the wing (mirrors) into vibrations to produce sound.
- Tympanum (Ears) are located on fore tibiae
- Tarsi are 4 segmented.
- Auditory organs are found in fore tibiae with a pair of tympanum is present on each fore tibiae.

Family: Gryllidae

Ex. Cricket

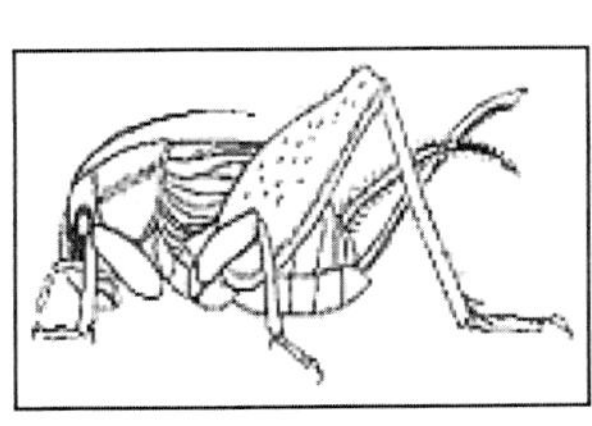

- Long antennae
- Tarsi are 3 segmented.
- Globular heads, long cerci, wings fit like a lid
- Females have needle-like ovipositors
- Forewings are broader across the top than Tettigoniidae and forewings are abruptly bent down to cover the sides of the body

- Hind wings are acuminate. They are produced into a pair of Long processes which project beyond the abdomen.
- Stridulate by rubbing bases of forewings
- Ears are found on outer and inner sides of the tibiae
- Cerci are long and unsegmented.
- Auditory organs and stridulatory organs are similar to long horned grasshopper. Males stridulate during night. They produce a shrill chirping noise. *Gryllus sp.* - household pets.

Family: Gryllotalpidae

Ex. Mole crickets

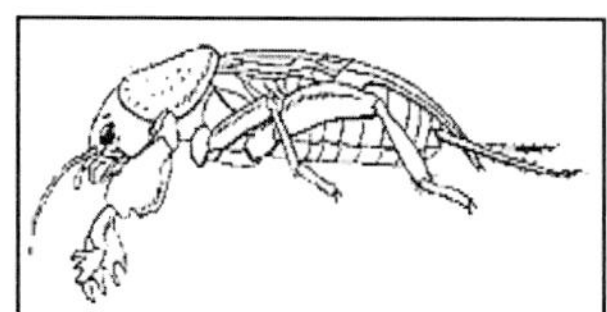

- They are brown coloured insects **found** inside the burrows.
- Eyes are reduced.
- Pronotum is elongate, ovate and **rounded** posteriorly.

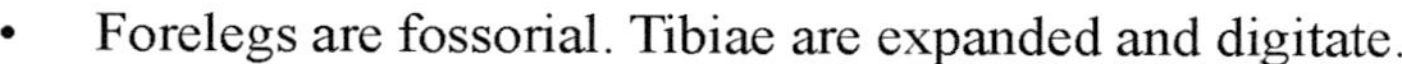

- Forelegs are fossorial. Tibiae are expanded and digitate.
- Hind wings are extended beyond the tegmina as a pair of processes
- Special stridulatory structure is absent. A humming sound is produced by rubbing theforewings.
- A pair of tympanum is found on the outer surface of the tibiae.
- Ovipositors vestigial.
- Mole crickets burrow into the soil and feed on tender roots of growing plants. *Gryllotalpa africana* is a pest on stored potatoes.

5.8 Phasmotodea (Phasma- phantom or apparition)

Ex. Stick insects, Leaf insects

Habitats: They are phytophagous and predominantly resembles (mimic) with various vegetation feature such as stems, stick and leaf. Amongst the foliage in which they are best camouflaged

Characteristic features

- Stick-like or leaf like elongate bodies and legs
- Large apterous or winged insect and frequently leaf like or elongated
- Wing pads do not undergo reversal during development. Head small prognathous
- Mouth part biting type

- Prothorax short, meso and metathorax usually elongated, they closely associated with 1st abdominal segment.
- Legs similar to each other. In leaf insect (Phyllidae) tibiae with a small triangular area whereas in stick insect (Phasmatidae) tibiae without a triangular area.
- Autotomy (shedding of appendages) is present in some species of stick insect as a defensive strategy against predators.

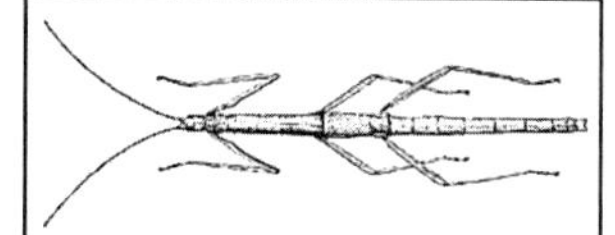

- Tarsi 5 segmented.
- Cerci short and unsegmented.
- Specilised stridulatory and auditory organ absent.
- Eggs deposited singly and need 1-2 years to hatch.
- Metamorphosis slight hemimetabolous type.

Important family

Phasmatidae (stick insect)

Phyllidae (leaf insect)

5.9 Mantophasmatodea (Combination of the order names for Mantodea and Phasmatodea)

Ex. Gladiators / Heelwalkers

Habitat: They are found in moderately humid to very dry habitats, where they live on small bushes or in the tussocks of grass-like plants. It is an amalgamation of the order names for Mantodea and Phasmatodea. So that the blend of physical and ecological characteristics found in these insects.

Characteristic features

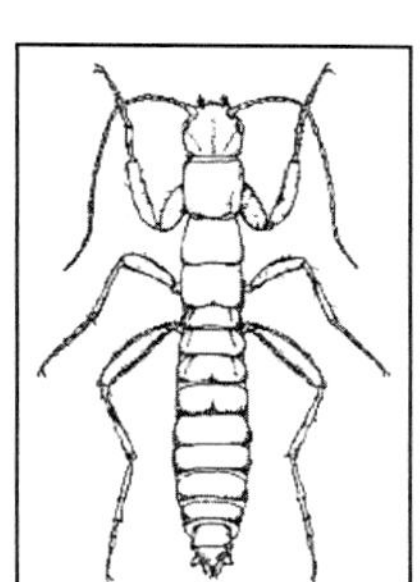

- New order 1st described in 2001
- Small, generally 2-3 cm length.
- Both sexes are wingless.
- Mouth part chewing type.
- Head hypognathous
- Antennae long and filiform.
- Tarsi 5-segmented
- Cerci short and one segmented. Resemble to mantid but fore leg no modified to prey capture.
- Simple metamorphosis.

5.10. Embioptera (Embio - lively, Ptera - wing)

Ex. Web spinners, Embiids

Habitat: All species are gregarious, with nymphs and adults occupying silken galleries. Species feed on bark, moss, lichen, dead leaves, and other plant material. Embioptera are the only insects which produce silk as adults, generating it in the glands in their enlarged front tarsi.

Characteristic features

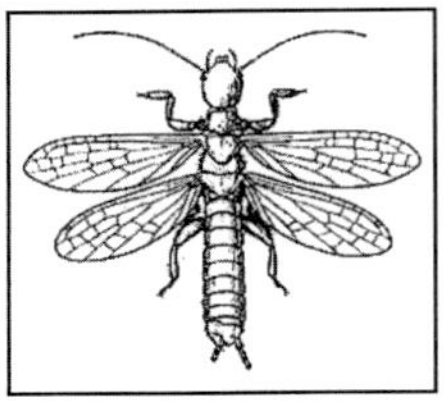

- Lively wing refer to fluttery movement of wing. Web spinners are elongate, mostly small insects
- Head prognathous type with chewing mouthparts.
- The antennae are filiform
- Ocelli are absent.
- Females are wingless; males are winged or wingless, and sometimes both conditions occur in males of one species. When wings are present, the wing venation is reduced.
- Legs contain tarsomeres.
- Males, females, and all instars spin silk from glands located on the fore basitarsus.
- The silk is secreted from hollow, tubular structures on the ventral surface of the basitarsus and second tarsomere.
- The abdomen appears 10-segmented, but the 11th segment is reduced.
- Cerci two-segmented (generally asymmetrical in the male) and tactile.
- Cannabalism may occur, and males are sometimes eaten by females after copulation.
- Rapid rearward movement of webspinners is enabled through large depressor muscles in the hind tibiae.
- Eggs and early instar nymphs are guarded by the female.

5.11. Zoraptera (Zor-pure, Aptera- wingless)

Ex. Angel insects

Habitats: The zorapteran are very small order of insect and mostly tropical. Development is gregarious, usually under planks, in piles of old sawdust, in rotting logs, under bark, or in association with termites. Zorapterans are apparently fungivorous or necrophilous or both.

Characteristic features

- Small, soft bodied termite like orthopteroids and usually wingless.
- Wings when present membranous and with few veins
- Fore wings larger than hind wings.
- Wingless forms lack of ocelli and compound eye while winged forms have compound eye and 3 ocelli.
- Haed prognathous
- Antennae moniliform
- Mouth parts biting type
- Prothorax is larger and globular.
- Abdomen short,oval and 10-segmented
- Tarsi 2-segmented
- Cerci small and unjointed
- Ovipositor absent

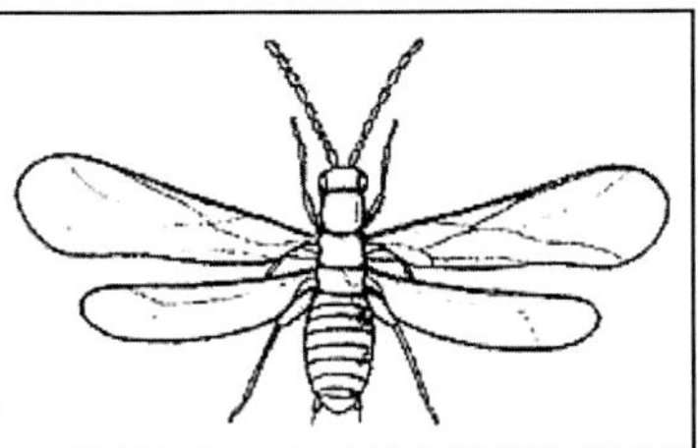

6

Paraneoptera

Common characters

- Modification of mandiblate mouth parts into sucking-piercing or suctorial type.
- Absence of anal lobe in hind wing.
- Presence of minimum numbers of malpighian tubules.
- Condensation of ventral nerve.
- Cerci absent.

6.1. Order: Psocoptera (Psokos - Rubbed or gnawer; ptera - wing)

Ex. Book lice, Bark lice, Dust lice, Psocids

Habitats: They are found in vegetation and under bark, occasionally on books Feed on starchy substances, cereal plants, mould (on books). Psocoptera can easily be confused with the small psyllids (Hemiptera) but the most obvious difference is their lack of hemipteroid beak. Several psocids live gregariously and some covered by a canopy of silk.

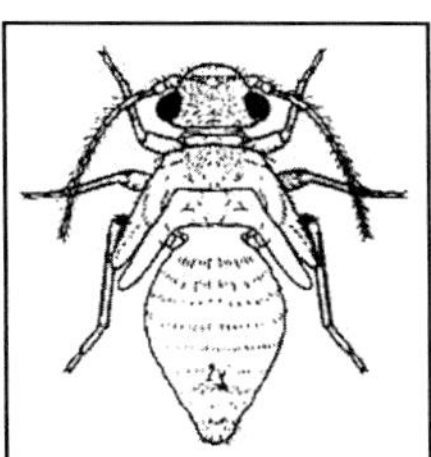

- Small (1-10 mm long) soft body insect with two pairs of membranous wings, or wingless.
- Long filiform antennae with 12-50 segments.
- Head with Y- shaped epicranial suture.
- Biting mouthparts
- Maxillae with 4 segmented palp and rod like lacina.
- Labial palp much reduces with 1 or 2 segmented.
- Tarsi 2 or 3 segment.
- Cerci absent
- Alary polymorphism is common. Eggs laid in group and cover with silken threads.

  Ex. Psocids, *Psocus* spp.

  Book louse, *Lipo scelis*

Difference between Book lice and Bark lice

Book lice	Bark lice
Bark lice generally live in moist terrestrial environment	More common in human dwelling and warehouse.
Winged during adult stage.	wingless
More than 10 mm length	Less than 2mm lenggth
Tarsi 2 segment	Tarsi 3 segment
Pearmans organs a sound producing organ present in the hind coxae of make that attract female	-

6.2. Order: Phthiraptera (Phthir –lice, aptera- wingless)

Ex. Biting and Sucking lice

Habitat: Parasite of animals and human being.

Characteristic feature

- Mouth part biting or piercing and sucking type.
- Small wingless insect.
- Eye reduces or absent.
- Ocelli absent.
- Antennae 3-5 segmented.
- Thoracic segment fused,
- Thoracic spircles ventral or dorsal.
- Ocelli absent.
- Permanently ectoparasitic on animal and mammals.
- Incomplete metamorphosis.
- Eggs of this order known as Nits.

This order divided into two sub order:

Sub order: Anoplura (Siphunculata)

Ex. Sucking lice

Important characters

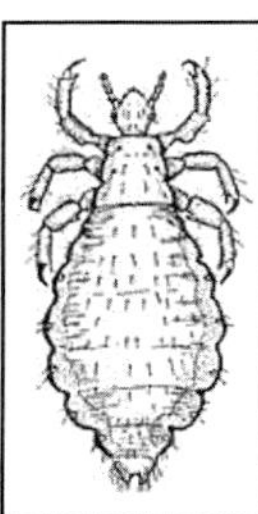

- Apterous insect living as ectoparasitic on mammals.
- 0.5-10 mm long, narrow head, pear-shaped, dorsoventrally flattened body.
- Eye reduce or absent and ocelli absent.
- Antennae 3-5 segmented.
- Thoracic segment fused tarsi one segment and single claw.

- Mouth part piercing and sucking with three styletes fused maxillae.
- Thoracic spiracle dorsal.
- Cerci absent.

Important family:

1. **Family:** Pediculidae

- Head with eye but with ocular regions. Antennae 5 segmented
- Clinging apparatus well devellop

Ex. *Pediculus humanus*

P.capitis (head louce)

P.corporis (body louse)

2. **Family:** Phthiridae

- Head with eye but no with ocular regions. It occurs on hairs of pubic region.

Ex. *Phthrius pubic* (pubic louse)

Sub order: Mallophaga

Ex. Biting or Bird lice

Important characters

- Apterous insect leaving as ectotoparasite on birds.
- Eye reduce and no ocelli
- Antennae 3-5 segments.
- Maxillary palp 4 segmented
- Labial palp rudimentary
- Tarsi 1-2 segmented and terminated by single or pair of claws
- Thoracic spiracile ventral
- Cerci absent.

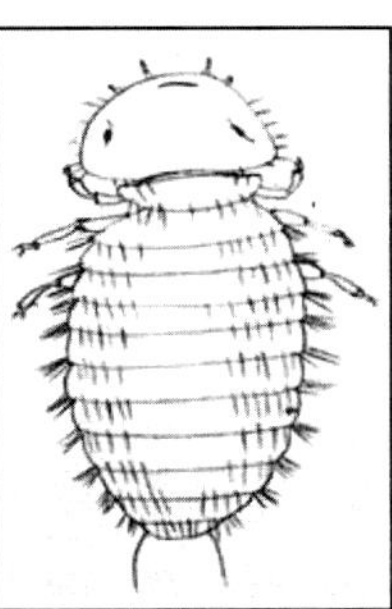

Defference between biting and sucking lice

Biting lice	Sucking lice
Body flattened	Oval
Mouth part biting type	Sucking type
Ectoparasite on birds	Ectoparasite on mammals
They feed feathers,skin and epidermal secreation	They feed on blood mainly
Thoracic spiracles ventral	Dorsal

6.3. Hemiptera (Derive from Latin word, hemipterus -half-winged)

Ex. True bugs/ Typical bugs

Habitat: Most hemipterans suck plant sap; some are predators or parasitic on mammals (Cimicidae and Polyctaenidae). Some species damage crops being vectors of viral diseases and some act as vector of pathogens causing human diseases (Reduviidae and Cimicidae). Other species have been used for biological control of insect pests. Hemipterans have been cultivated for the extraction of dyes and lac. Many species are used as food.

Characteristics

- Mouthparts piercing and sucking.
- Mandibles and maxillae modified into needle-like stylets, and labium form a rostrum or proboscis.
- Head mostly opisthognathous (true bug) and some cases hypognathous.
- Maxillary and labial palp absent.
- Pro- and mesothorax usually large, metathorax small
- 2 pairs of wings (some species wingless), venation reduced; male scale insects with one pair of wings.
- Forewings generally hardened to some extent.
- Cerci lacking

Difference between heteroptera and homoptera

S.No.	**Heteroptera** *Hetero-different; ptera-wing*	**Homoptera** *Homo-uniform; ptera-wing*
1.	Head is porrect or horizontal.	Head is deflexed
2.	Bases of the forelegs do not touch the head.	Bases of the forelegs touch the head.
3.	Beak arises from the anterior part of the head.	Beak arises from the posterior part of the head.
4.	Gular region of the head (midventral sclerotised part between labium and foramen magnum) well defined.	Gular region not clearly defined.
5.	Pronotum usually greatly enlarged.	Pronotum is almost alway small and collar-like.
6.	Scutellum (triangular plate, found between the wing bases) well developed	Scutellum not well developed.
7.	Forewings heavily sclerotized at the base and the apical half is membranous (Hemelytra)	Forewings are of uniform texture. They are frequently harder than hind pair.
8.	Wings are held flat over the back at rest and the left and right side overlap on the abdomen.	Wings are held roof-like over the backand wings do not overlap
9.	Honey dew secretion uncommon	Honey dew secretion common
10.	Repungnatorial or odori- ferous or scent glands present.	Wax glands usually present
11.	Both terrestrial and aquatic.	Terrestrial.
12.	Herbivorous, predaceous or blood sucking.	Herbivorous

It is divided into five **Suborders:**

1. Suborder: Sternorrhyncha
2. Suborder: Fulgoromorpha
3. Suborder: Cicadomorpha (Fulgomerpha and Cicadomorpha known as Auchenorrhyncha)

4. Suborder: Coleorrhyncha – (sister group of Heteroptera)
5. Suborder: Heteroptera

1. Suborder: Sternorrhyncha (aphids, whiteflies, scale insects)

- Antenna usually well developed without a terminal arista.
- Rostrum apparently sternum between fore coxae.
- Tarsi 1-2 segment.
- Many species of female and immature stage inactive or incapable to locomotion.

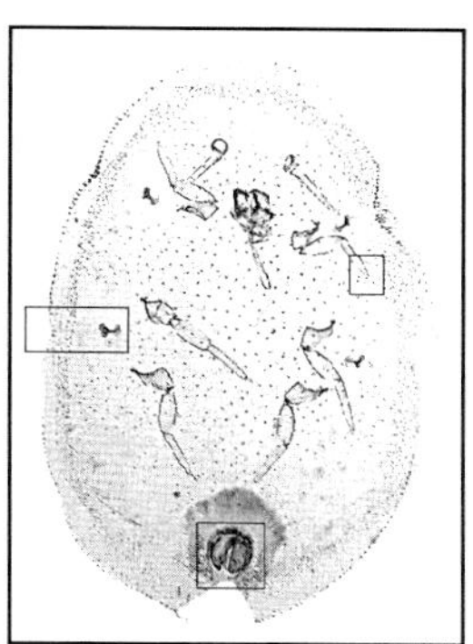

Family: Coccidae

Ex. Scale Insects, soft scales, wax scales

- Females are apterous, flate, elongated, oval, sack and gall like with with integument smooth or corved with wax.
- They are wingless, leg less and suck plant sap
- It is known as sessile insects that secrete a protective cover over soft bodies.
- Sexual dimorphism present.
- Male with long antennae, lateral eye and vestigial mouthparts.
- Mesothorax with a pair of wings with one or two veins.
- Hind wings reduced to halters.
- Tarsi 1 segmented with single claws.
- First instar nymph is active and known as crawler which moults and becomes leg less

Ex. Coffee green scale, *Coccus viridis*

Family: Lacciferidae/ Kerridae

Ex. lac insect

- Apterous Females with degenerated irregularly globular body, without legs and antennae.
- The body is irregular globose and enclosed in a thick resinous cell, secreted by lac resin gland/dermal gland.

Ex. Lac insect, *Kerria lacca*

Family: Psyllidae

Ex. Jumping plant lice

- They are very minute 2 to 5 mm, move actively by leaping and flying and feed on phloem sap. Some species form galls.
- Hind leg of femur thickened modified for leaping

- Tarsi 2 segmented with a paired of claws.
- 10 segmented antennae.
- Rostrum 3 segmented.
- A prominent basal vein in forewing formed by fusion of radius, median and cubitus

  Ex. Citrus psylla, *Psylla mali*

  Mango shoot, gall psylla, *Apsylla cistellata*

Family: Aleyrodidae

Ex. whiteflies

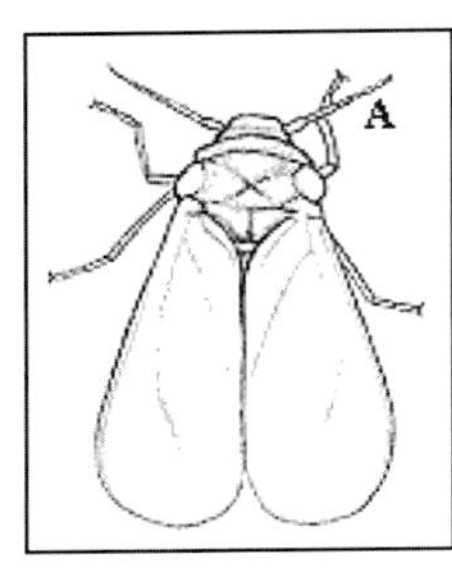

- Minute insect with 2 or 3 mm long and posses 7 segmented antennae, 2 ocelli.a pair of compound eye.
- The adult of both sexes are winged and the winges are cover whitish dust or waxy power.
- Rostrum 3 segmented and tarsi 2 segmented.
- Vasiform orifice (honeydew secreting organ which consist of operculum and lingula) open on the dorsal surface of last abdomin2al segments. Insects are oviparous
- Ist instar nymph are mobile and next 2 instar are immobile and 4th instar called pupa which is immobile.
- Homometabolous due to presence of a quiescent stage prior to adult emergence

  Ex. Cotton whitefly, *Bemisia tabaci* (transmits vein clearing disease in *bhindi*)

Family: Aphididae

Ex. aphids, green flies, plant lice

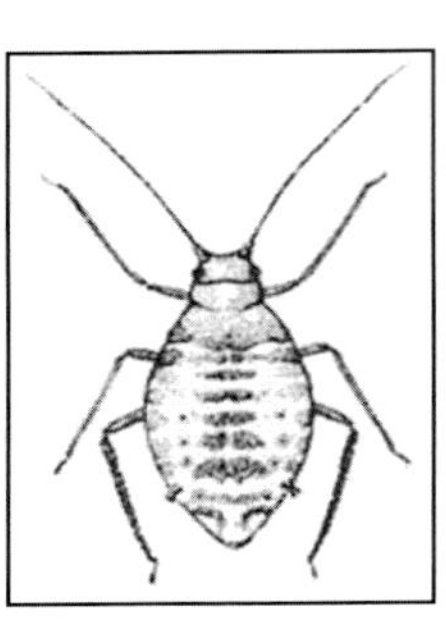

- Pear shaped body with pair of compound eye.
- Winged or wingless Adults.
- Antennae 2 to 3 segmented and Ocelli absent.
- A pair of cornicles present at poseterior end of 5th or 6th abdomeninal segment which secretes wax like substance for defense.
- A pair of lateral spiracle is present.
- Tarsi 2 segmented with a pair of claws.
- Excrete copious amount of honey dew through anus on which ants feed and sooty mould fungus grows.
- Fecundity high, life cycle short, parthenogenetic.

- Life cycle involves alternation of generation.
- Feed on plant sap and disseminate plant diseases
  Ex. Cotton aphid, *Aphis gossypii*
  Mustard aphid, *Lipaphis erysimi*

Family: Pseudococcidae

Ex. mealy bug

- Females, body elongate-oval segmented
- Mealy bug body is covered by a white sticky powder
- The females and "crawlers," or active young, cluster along the veins on the undersides of leaves
- Males are active fliers and have only two wings.
- Have well developed legs

2. Suborder: Fulgoromorpha

Family: Delphacidae

Ex. planthoppers

- Hind tibia with apical large mobile, flattened spur.
- Ovipositor fully developed.
- Antennaal pedical with large and numerous sensilla.
- Feed mainly on monocots like rice, wheat, corn, and sugarcane.
- Many of them cause viral diseases by direct feeding
  Ex. a. Brown planthopper, *Nilaparvata lugens*
  - Causes hopper burn
  - It transmits viral disease in rice

  b. White-backed rice planthopper, *Sogatella furcifera*

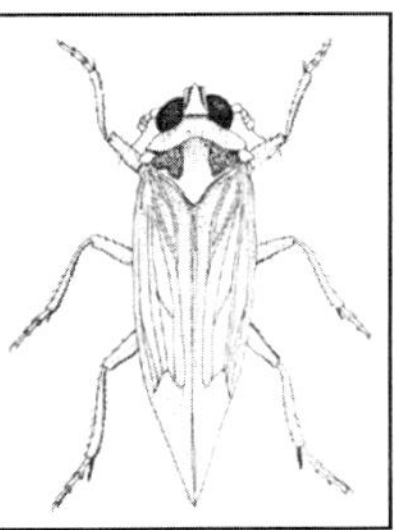

Family: Lophopidae

Ex. pyrilla

- Vertex with width less than 3 times of length
- Head prolonged
- 1-3 longitudinal carinae

3. Suborder: Cicadomorpha

Ex. cicadas, spittlebugs, leafhoppers and treehoppers

Family: Cicadellidae

Ex. leaf hoppers or jassids

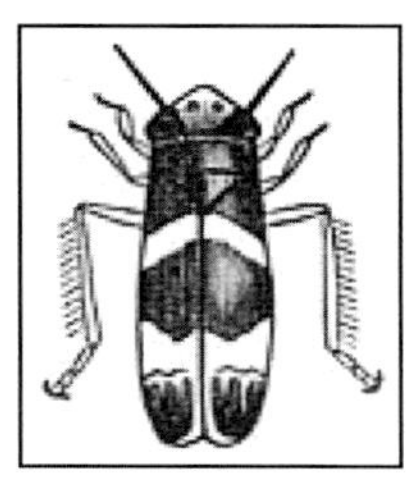

- Elongate, coloured insects
- They have one or more rows of small spines extending the length of hind tibiae.
- Ovipositor well suited for lacerating plant tissue.
- Suck plant sap and transmit diseases

  Ex. Green leaf hopper, *Nephotettix virescens* (transmits tungro disease in rice)

  Mango leaf hopper, *Idiocerus atkinsoni*

Family: Cicadidae

Ex. cicadas

- Female without timbals.
- Males have sound producing organs at base of abdomen called Timbal.
- Each species has a characteristic song.
- Tympanum present in both sexes
- Wings transparent.
- Nymphs drop to the ground, enter soil and feed on root sap.
- Anterior femora of nymph thickened with spines beneath and suited for digging soil
- Life cycle of periodical cicada lasts for 13-17 years

  Ex. *Platypleura* spp.

Family: Membracidae

Ex. treehoppers

- Pronotum is pointed projecting backward over abdomen. Treehopper feed on trees and rubs.
- Stridulatory organ is absent. *Tricentrus* spp. is a pest on crops.

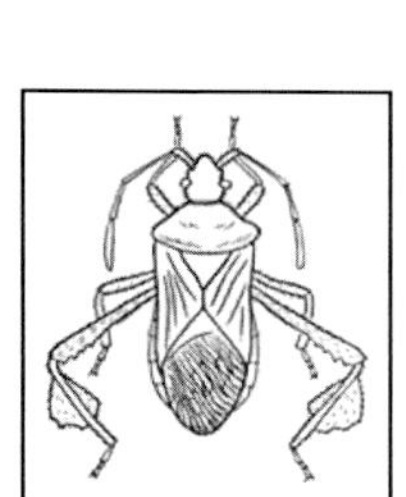

4. Sub-order: Hereteroptera

Family: Coreidae

Ex. squash bugs and leaf-footed bugs

- Head narrower than thorax, ocelli present
- Many parallel veins in membrane of fore wings
- Hind tibia expanded into thin, leaf-like plates

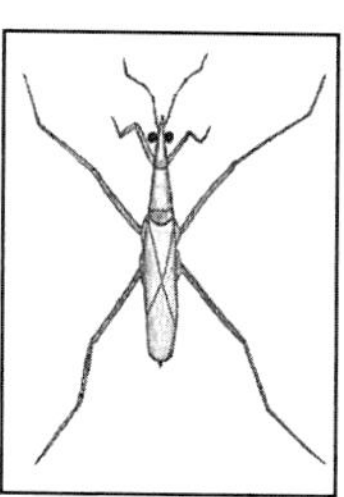

- Scent gland present between middle and hind coxae which emit a offensive odour
- Nymphs and adults suck sap from pods of pulses

Ex. *Cletus* spp.

Family: Gerridae

Ex. water striders, pond skater

- Forelegs short, raptrorial and suited for capturing prey.
- Mid legs long and useful in pushing.
- Hind legs long and help in steering.
- Legs with fine non wetting hairs.
- Skate on water surface.
- Feed on insects falling on water surface

Ex. Water strider, *Gerris* spp.

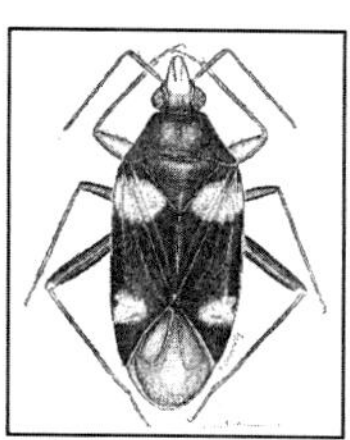

Family: Miridae

Ex. leaf bugs, plant bugs

- Beak and antennae, 4 segmented.
- Ocelli absent.
- Hemelytra with distinct corium, clavus and cuneus (a triangular apical piece of the basal part of forewing). Loop veins found in wing membrane.
- Nymphs and adults feed on plant juice and some species inject toxic saliva in plants.
- A few species are predaceous

Ex. sorghum earhead bug, *Calocoris angustatus*

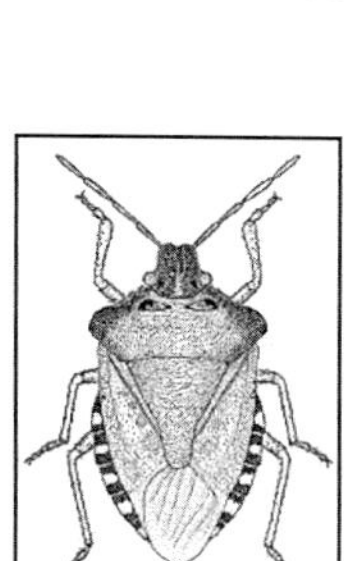

Family: Pentatomidae

Ex. Stink bugs

- Shield-shaped body.
- 5 segmented antennae.
- Large triangular scutellum in center of back.
- Adults and nymphs produce a odour from stink glands located in metathorax and abdomen.
- Phytophagous; some species predaceous

Ex. Southern *green stink bug, Nezara viridula*

Predatory bug, *Eocanthecona furcellata.* Painted bug, *Bagrada hilaris*

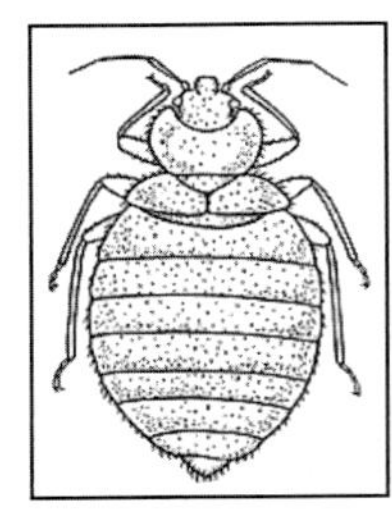

Family: Cimicidae

Ex. bed bugs

- Body flat, oval, dull reddish-brown can hide in cracks and crevices.
- Hemelytra short and reduced to scale like pads.
- Hind wings completely atrophied.
- Stink glands in dorsal surface of first three abdominal segments.
- Male bed bugs pierce integument of female and inject the sperm into haemocoel during copulation (haemocoelic or traumatic insemination).
- Blood sucking ectoparasites on birds and mammals

Ex. Common bed bug, *Cimex lectularius*

Cimex hemipterus (important species affecting man)

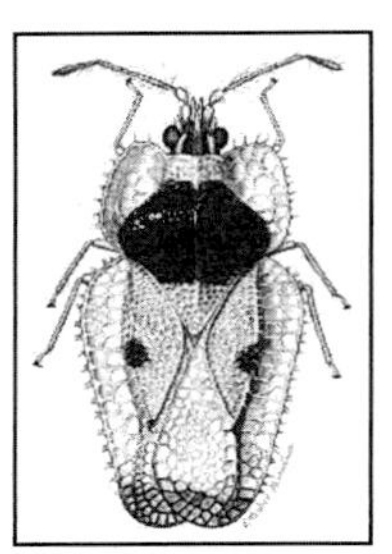

Family: Tingidae

Ex. lacewing bugs

- Pronotum laterally expanded with lace like sculpturing.
- Forewings with elaborate lace-like markings due to densely reticulate, raised wing venation.
- Nymphs usually spiny and lack lace-like markings.
- Both nymphs and adults found on undersurface of leaves in groups, suck sap and produce white spotted appearance on leaves.
- Secrete honey dew

Ex. *Tingis* spp.

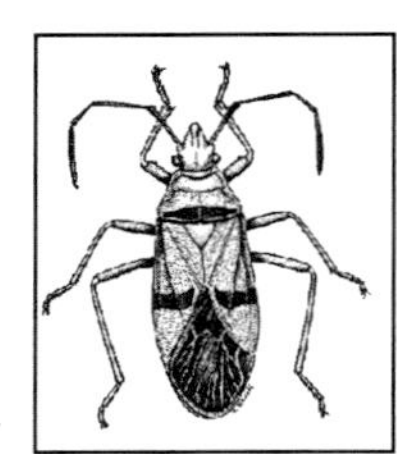

Family: Pyrrhocoridae

Ex. red bugs or stainers

- Elongate, oval bugs, brightly marked with red and black.
- Show warning colouration.
- Wings with branched veins and cells.
- Feeding injury cause contamination by fungus

Ex. Red cotton bug, *Dysdercus cingulatus*

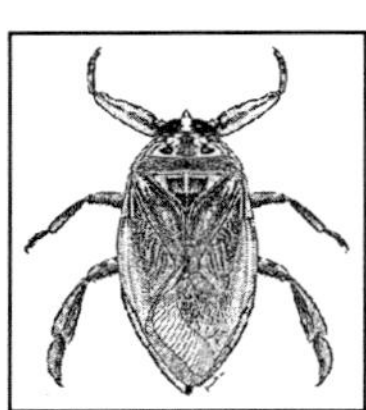

Family: Belostomatidae

Ex. giant water bugs or electric light bugs

- Large sized.

- Forelegs raptorial and suited for capturing prey.
- Posterior legs adapted for swimming.
- Tibia and tarsus flat and fringed with hairs.
- Abdomen with 2 short terminal breathing tube.
- Excellent fliers and swimmers.
- In some species eggs are laid on back of male.
- Suck blood from toads, frogs, fishes and even human beings
  Ex. *Belostoma* spp.

Family: Reduviidae

- Kissing bug/ Cone nose bug
- Body generally elongate oval
- Beak short 3-segmented
- Highly predaceous
- Kissing bug transmit chagas (American trypanosomiasis) disease and also bite humans and other vertebrate eg: *Rhodinus prolixus* for blood meal.

Family: Anthocoridae

- Also called minute pirate bugs
- Hemelytra with a cuneus and embolium
- Ocelli present
- Beak 3-segmented
- Commonly present on flowers and fruit and feed on eggs and small insects.

Family: Lygaeidae

Ex. dusky cotton bug

- Hemelytra with 4-5 veins
- Scent gland present
- Many species are pests of crop

6.4 Thysanoptera (Physopoda) (Thysano - Fringed / Physopoda-bladder footed and pteron- wing)

Eg. Thrips

Habitats: Most of the thrips species belong to the family Thripidae and are phytophagous. They suck the plant sap. Some are vectors of plant diseases. Few are predators.

Characterstic features

- Minute slender bodied terrestrial insects and phytophagous insect
- Antennae short moniliform, 6-10 segmented, usually with sense cones or sensoria on 3rd or 4th segments

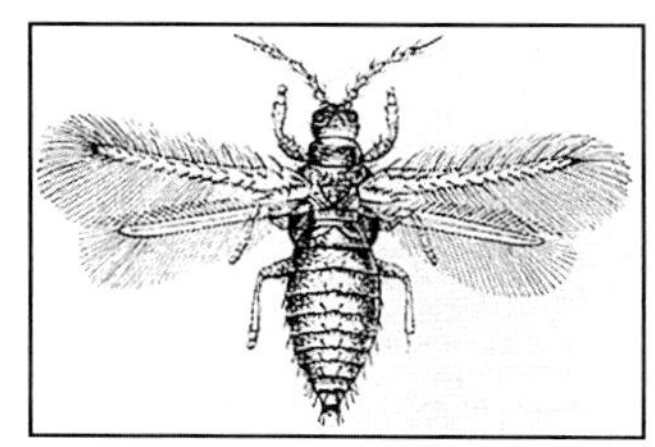

- Compound eyes conspicuous with 3 ocelli in winged forms
- Mouth parts asymenetrical, right mandible is rudimentary, lacerating and sucking or rasping and sucking type with three stylets.
- Mouth cone is formed by the labrum, labium and the maxillae which extend ventrally between the front coxae.
- Winged or wingless. Wings when fully developed are long and narrow with highly reduced venation (with few or no veins). The wings are fringed with long hairs on the margins.
- Legs short, tarsi 1 or 2 segmented with 1 or 2 claws, with a bladder like terminal protrusable vesicle.
- Abdomen is elongate with 10-11 segments, usually tapering posteriorly
- Cerci absent
- Metamorphosis is accompanied by one or two inactive pupal instars i.e intermediary between simple and complete.
- Parthenogenetic type of reproduction is very common and in many species males are rarely seen.
- E.g. Rice thrips: *Stenchaetothrips biformis* is a pest in rice nursery.

Sub order

1. Terebrantia

- Female bearing saw like serrated ovipositors
- Female with conical or round last abdominal segment
- Male has blunt and round abdominal apex
- Fore wing have veins and seatae

Family: Thripidae

- This is the largest and most important injurious family in thrips.
- Antennae 6-9 segments not conical with sense cones. Antennae ends in 1 or segmented apical style. 4th segment usually enlarged.
- Winged narrow and pointed at the tip and fringed with long hairs on the margins.
- Ovipositor curved downard

- Tarsi sometimes with claw like appendages at apex of 1st or 2nd segment.

Eg: Onion thrips -*Thrips tabaci*

Chilli thrips - *Scirtothrips dorsalis*

2. Tublifera

- Female whitout ovipositors
- Last abdominal segment tubular
- Fore wing lack of veins and seatae

Family: Phlaeothripidae

- Antennae with 7-8segment
- 3rd antennal segment is the largest one.
- Head rounded
- Maxillary palp 2 segmented

Eg. Phlaeothrips

7

Endopterygota (Holometabola)

- Wings develop internally
- Metamorphosis complete
- Immature stages (larvae) differ from adults in structure, habitat and feeding habits
- Pupal stage present, e.g. all holometabolus insects

7.1. Order: Neuroptera (Neuro-nerve and Ptera-wing)

Ex. Lacewings, Mantidflies, Antlions

Habitat: Neuropteran larvae are carnivorous and free-living with the exception of the aquatic family Sisyridae (spongillafiles), which has larvae that are parasitic on freshwater sponges. Typically, a neuropteran larva sucks out the contents of its prey, leaving only a hollow skin. Although many lacewing larvae are nocturnal and need no camouflage, other species carry debris on bodies adapted for this purpose. They also serve as food for fish and other aquatic vertebrates. Lacewing larvae are beneficial as predators of agricultural pests (aphids, whiteflies and scale insects). Some species are reared and sold commercially as biocontrol agents.

Characteristics features

- Neuropteran is very closed to order Megaloptera and Raphidioptera
- Wings two pairs with extensive branch so called nerve wing
- Wings membranous narrow at base, net-veined, equal or subequal; held roof-like over abdomen when at rest
- Soft-bodied insects; large lateral compound eyes; ocelli present or absent
- Mouth part biting and chewing type
- Antennae filiform, multisegmented.
- Head prognathous
- Cerci present
- May show complex courtship and mating rituals

Immatures

- Head well-developed with ocelli, antennae.
- The larvae have specialized mandible and maxillae. They interlocked to form pincer and with the help of pincer suck the body of host.
- Some species of larvae is aquatic in nature and they respire through gills on most abdominal segments.
- Larvae predators and suck prey's body contents through hollow food channels

  Larvae campodeiform
- Larvae are important predators of agricultural pests (aphids, whiteflies and scale insects
- Most of species spin silken cocoons.
- Some species feed on ants known as antilions.

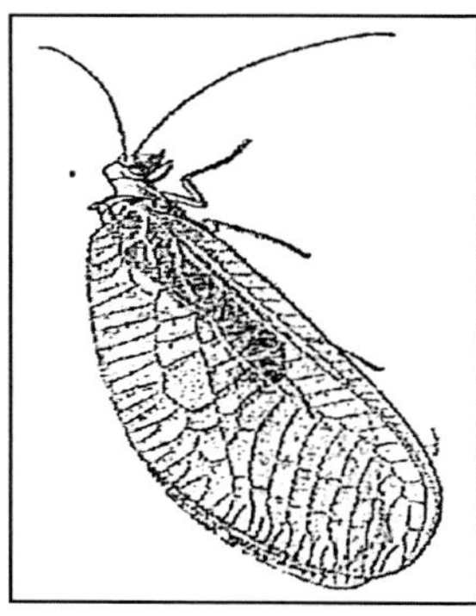

Lacewing

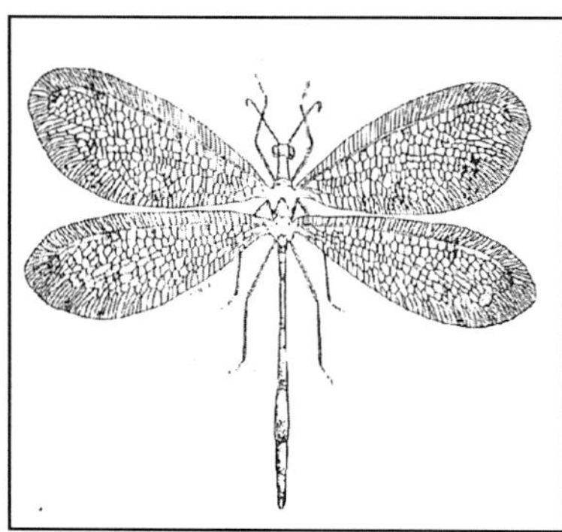

Antilon

Important families

Chrysopidae

- Antennae filiform, larger than fore wing
- Mandible long and strong
- Wings and body are green with golden eye
- Trichosors absent
- Larvae with hooked hair on the body for supporting dead host
- Eggs stalked

  Empodium trumpet shaped

Myremeleonitdae: Ex.Antlion

Ascalphidae: Ex. Owlflies

Mantispidae: Ex. Mantidfly

7.2. Order: Raphidioptera (raphid = needle, and ptera = wings)

Ex. Snakeflies

Habitat: Adults can be found in early spring, and fly only a short distance. They feed actively on small bodied insects, including their own congeners if put together. They are also reported to take nectar.The elongate larvae are terrestrial, with mandibulate, chewing mouthparts. They live in bark crevices, or in areas with dead leaves or debris. They are aggressive predators, feeding on small arthropods. Pupation takes place in dead leaves, debris or crevices. There is no cocoon and the pupa is free (exarate), active and able to use its mandibles.

Characteristics features

- Prothorax of adult Snake fly is elongated and gives appearance of snake.
- Male and female snake fly are quite distinct from each other.
- The female have very long ovipositors and look like tails or stinger.
- Head prognathous, mouthparts, chewing.
- Large compound eyes; some species with ocelli
- Long, thin antennae

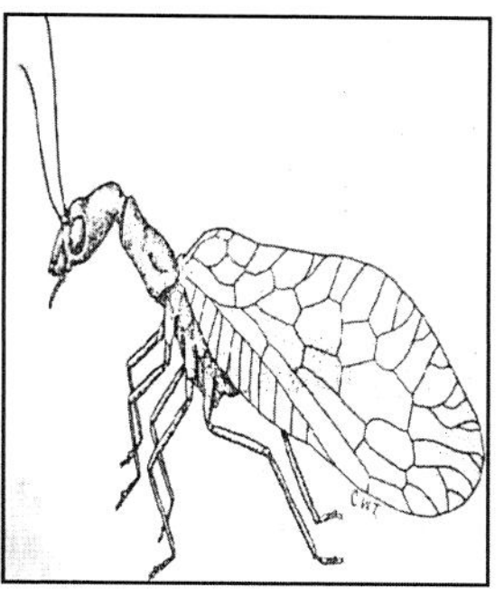

Snakefly

- Wings similar, venation primitive, with a thickened costal margin (or 'pterostigma'); held roof-like over abdomen when at rest.
- Females with long ovipositor, which is used to deposit eggs into crevices in bark or rotting wood.
- Larvae terrestrial, with chewing mouthparts; without prolegs; do not spin cocoon.
- Both larvae and adults are predator and feed on soft-bodied insects.

  Ex. *Raphidia* spp.

7.3. Order: Megaloptera (Mega-large and Ptera-wing)

Ex. Dobsonflies, Alderflies and Fishflies

Habitat: Adults have a slow, clumsy flight, and either do not feed or at most feed on small quantities of nectar or fruit juices.Larvae are aquatic and predaceous with biting mouthparts, a distinct labrum and maxillary palps. They are elongate, with characteristic lateral abdominal filaments, but they do not spin cocoons.

Characteristics features

- Medium to large, soft-bodied insect. Some adults are also very large in size and wing span about 16 cm.
- Head prognathous.
- Long antennae filliform.
- Two pairs of membranous wing which are sub equal in size with dark patches with many veins and crossveins and at rest wings are held roof-like over body.
- Hind wings broader than fore wings and with a large anal area.
- Chewing type mouth part with strong mandible.
- Compound eye large in size and ocelli absent
- Larvae are aquatic and largest in size among all aquatic insect.

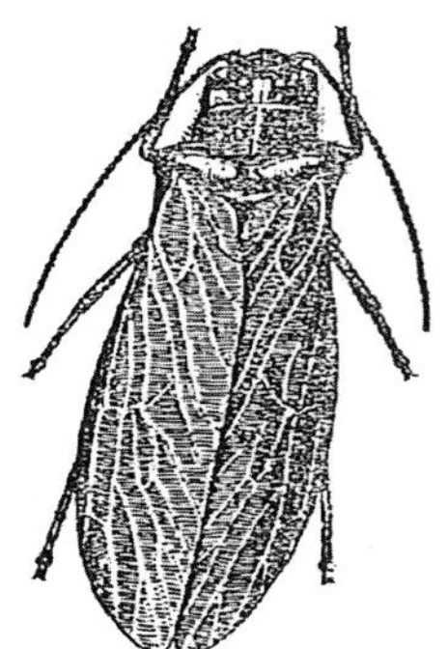
Alderfly

- The larvae have gills and tactile filament which are sensory in nature.
- Final segment of abdomen of larvae bears either a pair of prolegs, or a single, tail-like appendage
- They are predatory in nature and feed on aquatic insect and other animal.
- Ex. *Sialis* sp.

7.4. Order: Coleoptera (Coleo - sheath, pteron - wing)

Ex. Beetles and weevils

Habitat: They feed on a wide variety of diets, inhabit all terrestrial and fresh-water environments, and exhibit a number of different life styles. Most beetles are imporannt pests of agricultural plants and stored products. They are herbivores and variously adapted to feed on the roots, stems, leaves, or reproductive structures of their host plants. Some species also live on fungi, others burrow into plant tissues, still others excavate tunnels in wood or under bark. They are predators also and some beetles are scavengers, feeding primarily on carrion, fecal material, decaying wood, or other dead organic matter. Beetles live in the soil or on vegetation and attack a wide variety of invertebrate hosts. There are even a few parasitic beetles some are internal parasites of other insects, some invade the nests of ants or termites, and some are external parasites of mammals.

Characteristic features

- Elytra are sheath like (fore wings)
- It is the largest order in class insecta comprising about 35-40% of the known insect species.
- Minute to large sized insect with leathery or horny integument.
- Head highly sclerotized, free, normal or prolonged in to a snout as in weevils.
- Ocelli usually absent.
- Antennae variable usually 11 segmented
- Chewing mouthparts (sometimes located at the tip of a beak or snout) with well developed mandibles. The mandibles attain their greatest length in the males of many of the stag beetles (Lucanidae)
- Prothorax large and freely movable, mesothorax much reduced and fused with metathorax and the tergum of these segments is divisible in to prescutum, scutum and scutellum.
- Two pairs of wings present, front wings modified to form elytra.This structure is hard and serve as covers for the hind wings; meet in a line down the middle of the back
- Hind wings large, membranous and folded beneath the elytra
- Legs well developed for walking and running
- Tarsi 2- to 5-segmented
- Abdomen usually 10 segmented. First tergum membranous and one or more of the sterna from the first to third are aborted in many species, the terminal abdominal segments are refractile and tubular, thus functioning as an ovipositor (eg: cerambycidae).
- The larvae are known as grubs and generally thoracic legs are present.
- Pupa exarate, pale coloured and are invested by a thin soft cuticle.
- Most of the adults possess stridulatory organs and these are variable.
- Metamorphosis is holometabola i.e. complete metamorphosis (egg, larva, pupa, adult).

Coleoptera is divided into four suborders : Myxophaga, Archostemata, Adephaga and Polyphaga but only two of these, Aedephaga and Polyphaga, include agricultural important families:

Aedephaga	Polyphaga
• The first abdominal sternum is divided by the hind coxae and the posterior margin of this sternum do not extended completely across the abdomen.	• The first abdominal sternum is undivided and its posterior margin extended completely across the abdomen.
• Hind trochanters are usually large	• Hind trochanters are usually small
• They have notoplural suture	• Notoplural suture are lacking
• Tarsal formula 5-5-5	• Tarsal formula Variable

ADEPHAGA

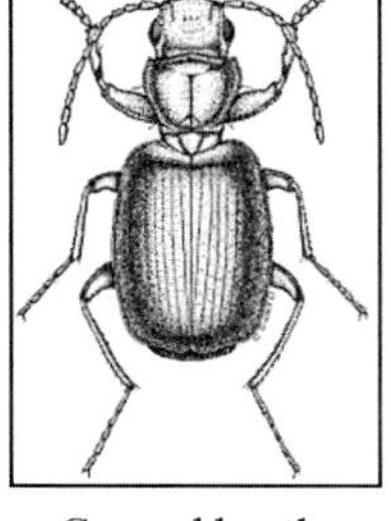
Ground beetle

Family Carabidae

Ex. Ground beetles, Carabids

- Adults often black, some brightly spotted
- Mandibles large, toothed
- Large eyes
- Spiny powerful legs (suited for running)
- Some cannot fly because they have fused elytra and atrophied hind wings
- Nocturnal, Many beneficial, voracious predators both as adults and larvae, feed on soft bodied caterpillars, soil dwelling insects, ants, aphids and slugs
- Ex. *Carabus* spp., *Cicindela* spp.

Family: Dytiscidae

Ex.True water beetles, Predaceous diving beetles

- Body long, oval, smooth and shiny, with joined head, thorax and abdomen
- Antenna filiform
- Biting mouthparts adapted for predatory habit
- Hind legs flat, fringed with hairs and suited for swimming
- Air is stored beneath elytra
- Female oviposit in stems of aquatic plants

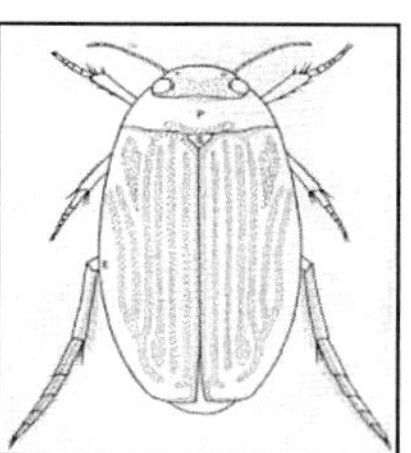

Ex. *Dytiscus* spp.;

Cybister spp.

Family: Cicindelidae

(Tiger beetles)

- Head is usually wider than prothorax.
- Eyes are fairly larger and they have very keen vision.
- Mandibles are sharply pointed, sickle shaped and acutely toothed for capturing the prey.
- Legs are long and tarsi slender which enable to run fast.
- Elytra have spots and stripes.
- Larva excavates vertical pits for prey capture.
- Both grubs and adults are active predators.
- Ex.*Cicindella sexpunctata*

Family: Gyrinidae

(Whirligig beetles)

- Gregarious surface swimmers
- Antennae short and clubbed
- Front legs elongated and slender
- Both adult and larvae predaceous

  Ex. *Gyrinus marinus*

POLYPHAGA

Family: Anobiidae

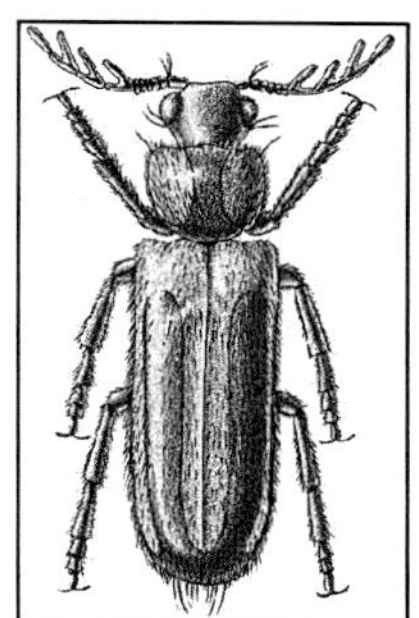

Ex. Anobiid beetles, Wood worms, Wood borers, death-watches

- Body oval or cylindrical
- 3 apical antennal segments large and free
- Head concealed by helmet-like prothorax
- Destructive to wood and stored products

  Ex. Cigarette beetle, *Lasioderma serricorne* (most serious pest of tobacco in factories and cigar stores)

Family Bostrichidae

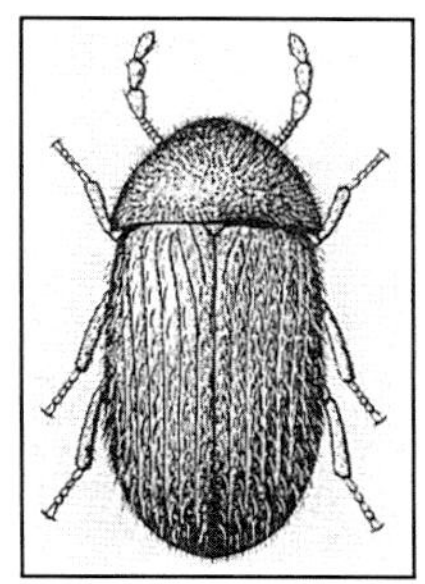

Ex. Branch and twig beetle

- Small, elongate and cylindrical, head concealed by hood-like pronotum
- Clubbed type antennae
- Elytra smooth or sculptured
- Burrow timber, dry wood, unhealthy trees

  Ex. Lesser grain borer, *Rhizopertha dominica*
- Larvae bore into stored grains and eat inner content

Family Lampyridae (Lightning bugs, Fireflies, Glow worms)

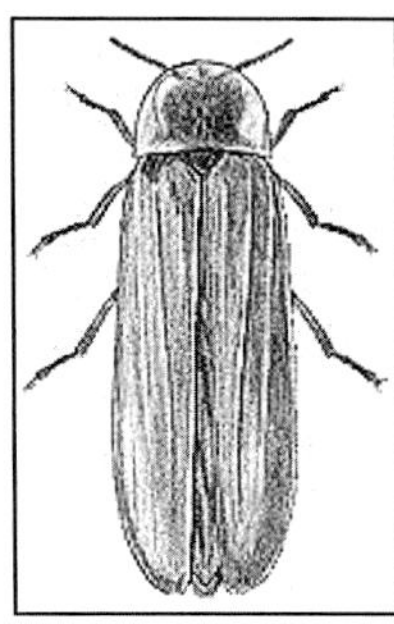

- Small to medium sized, elongated, luminous beetles
- Show sexual dimorphism. Male - with well developed eyes, winged and photogenic organ in 6th and 7th abdominal segments. Female – reduced eyes, wingless and larviform, photogenic organ in 7th abdominal segment
- Larvae carnivorous and feed on snails, slugs and earthworms
- All life stages are luminous to varying degree, luminescence is to bring both sexes together
- Ex. Common glow-worm, *Lampyris noctiluca*; firefly, *Luciola* spp.

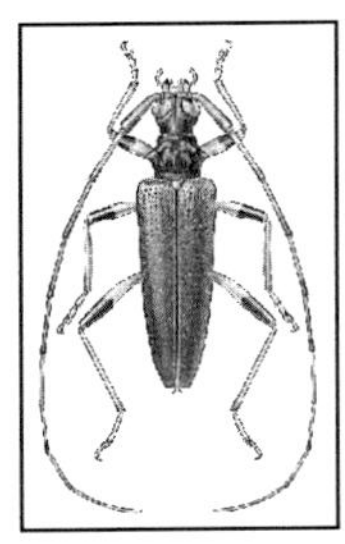

Family Cerambycidae

Ex. Long-horned beetles, Longicorn beetles

- Body cylindrical, brightly coloured
- Compound eyes notched
- Very long antenna
- Tarasus apparemtly 4-4-4 (but actually it is 5-5-5)
- Pronotum with 1-3 lateral spines
- Grubs called round headed borers, apodous, wood borers. They develop beneath bark and tunnel into branches or main stem

Ex. Mango stem borer, *Batocera rufomaculata*

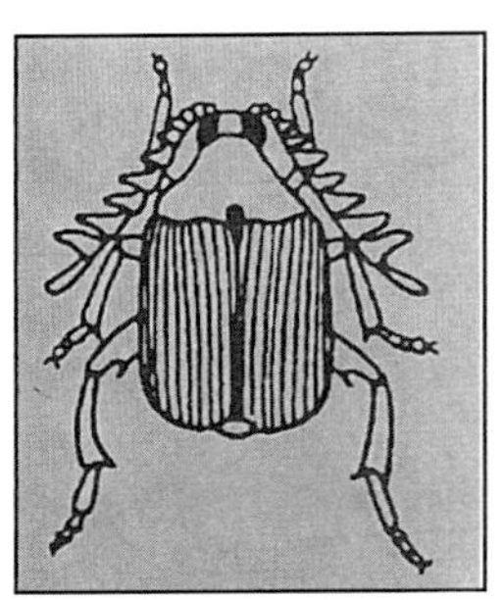

Family Bruchidae

Ex. Pulse beetles, Seed beetles

- Small with blunt snout; covered with scales
- Serrate type antenna
- Hind femur thick
- Elytra short and not cover abdomen fully (Pygidium)
- Eggs glued to pods or seeds by a glutinous secretion, grubs feed exclusively on seed legumes
- Adult and larvae feed inside seeds

EX. Pulse beetle, *Callosobruchus chinensis*

- It is important and serious pest on pulses of storage condition.

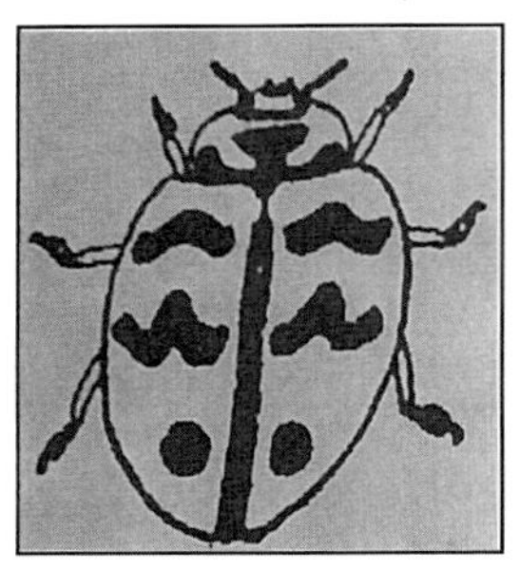

Family : Coccinellidae

Ex. Lady Bird beetles

- Hemispherical, body convex above and flat below.
- Head small, turned downward and received into a prominent notch of prothorax.
- Tarsi formula 4-4-4 but appear 3-3-3 & tarsal claws toothed at base
- Elytra strongly convex, brightly coloured and variously spotted.
- Grubs compodeiform and spiny
- Except genus *Epilachna* (phyophagous), others are predators on aphids, scales, mites and whiteflies (important bio-control agents)

Ex. *Coccinella septempuctata*

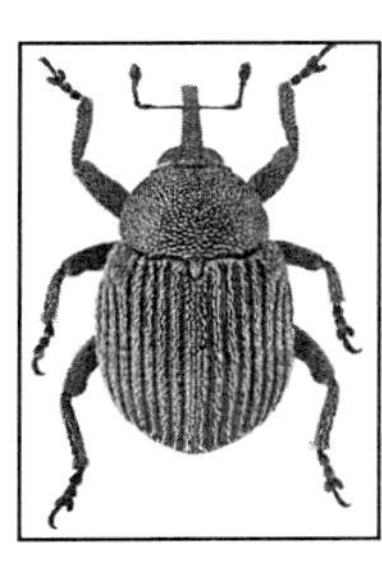

Family: Curculionidae

Ex. Weevils, Snout beetles, Bark beetles, Timber beetles

- Largest family
- Minute to large sized, head produced into snout.
- Mouthparts (Mandibles and maxillae) present at tip of snout and used to feed on internal plant tissues.
- Antenna geniculate.
- Grubs apodous and eucephalous.
- Weevils are important pest of crops & stored grains.
- Bark beetles mine under bark; timber beetles grow in galleries made in wood

Ex. Coconut red palm weevil, *Rhynchophorus ferrugineus*

Rice weevil, *Sitophilus oryzae*

Lucerne weevil, *Hypera postica*; *Xyleborus* spp.

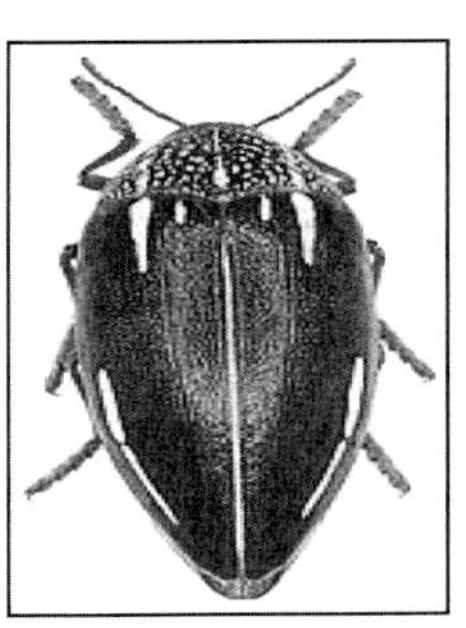

Family: Buprestidae

Ex. Metallic wood-boring beetles, Jewel beetles

- Elongate, hard bodied insects with metallic lustre.
- Antenna serrate.
- Tarsi 5-5-5 with lamellae beneath
- Larvae with small head, entirely withdrawn into thorax. Legs absent
- Herbivores, larvae tunnel beneath bark or bore into stems or roots
- Used in embroidery or jwellery

Ex. *Sternocera* spp.; *Trachys* spp.

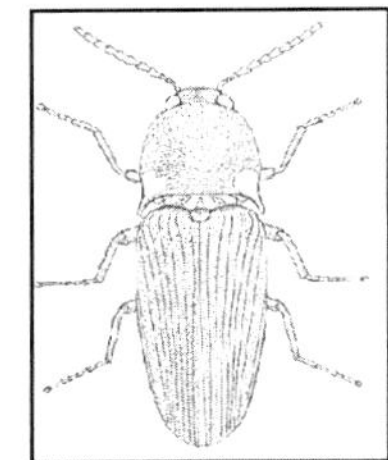

Family: Elateridae

Ex. Click beetles, Wire worms

- Body elongate, cylindrical.
- Pronotum rounded anteriorly.
- Prosternal spine work as "click" mechanism.
- Adult able to jump, each jump accompanied by an audible clicking sound.
- Herbivores, lay eggs in soil. Grubs long, cylindrical, flat and feed on roots

Ex. *Agrypnus* spp. *Elayter* spp.

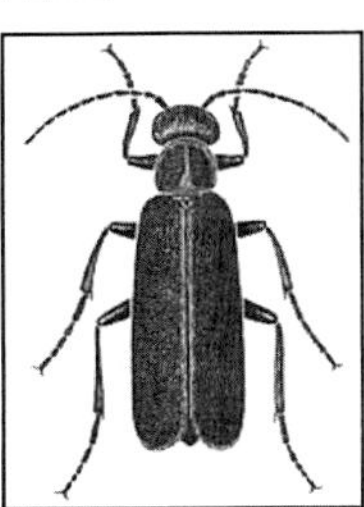

Family: Meloidae

Ex. Blister beetles, Oil beetles

- Cylindrical, soft bodied, covered with bright metallic colour; head connected to thorax by a distinct neck.

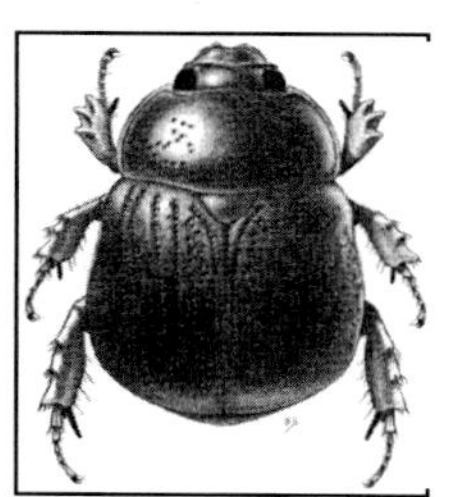

- Legs heteromerous with a tarsal formula of 5-5-4.
- Fore wings soft and leathery; wings vestigial or absent in some species.
- They give off a fluid containing oily principle *Cantharidin*, when disturbed which causes blisters.
- Development involves hypermetamorphosis (larval instars distinct from each other). Eggs hatch into active triungulin larvae which may feed on eggs of grasshoppers.
- Adults feed on foliage and flowers

  Ex. Orange banded blister beetle, *Mylabris pustulata*

Family: Scarabaeidae

Ex. Scarabs, Dung beetles

- Smooth, shiny, round or oval body
- Many with spines or horns on legs and prothorax.
- Head broad and flat, mandibles membranous and incapa
- Lamellate antennae
- Forelegs fossorial
- Adults and larvae scavengers, feed upon droppings of animals and human excreta. Make underground chambers and roll on dung into balls and bury them in; show remarkable parental care

  Ex. *Scarabaeus* sp.

Family : Tenebrionidae

Ex. Darkling beetles, Meal worms

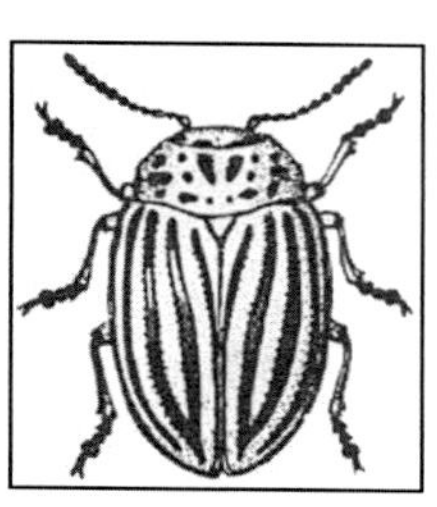

- Body flat, elongate
- Elytra often sculptured
- Many species apterous or with vestigial wings
- Legs heteromerous with a tarsal formula of 5-5-4
- Herbivores, feed on decaying vegetables, flour, dry seeds, and stored products; larvae called meal worms

  Ex. Red flour beetle, *Tribolium castaneum*
- It is important pest of milled products

Family : Chrysomelidae

Ex. Leaf beetles Tortoise beetles, Flea beetles

- Brightly coloured, oval to convex
- Eyes not prominent

- Antenna usually less than half length of body
- Tarsal formula apparently 4-4-4 , but actually 5-5-5 (as in case of cerambycid)
- Can hide their legs and antennae under wing covers while sitting on a leaf
- Herbivores; includes many pests of agricultural crops

  Ex. Colorado potato beetle, *Leptinotarsa decemlineata*; *Chrysomela* spp.

Family : Cantharidae

(Soldier beetles)

- Body covered with fine pubescence with soft integument
- Adult and larvae carnivorous

  Ex. *Cantharis fusca*

Family : Dermestidae

(Skin beetles, Carpet Beetles)

- Round oval elongated and body hairy or covered with scales
- Antennae short clubbed
- Pronotum strongly narrowed anteriorly
- Most species are scavengers and feed on dry plant and animal material and they can also damage carpets, draperies, and other woolen fabrics.

  Ex. Khappara beetle, *Trogoderma granarium*

7.5. Order Trichoptera (Trico-hair;ptera-wing)

Ex. Caddisflies, Sedge-flies or Rail-flies

Habitat: This order is another likely descendant of the mecopteran lineage. Adults are mostly nocturnal, weak-flying insects that are often attracted to lights. During the day, they hide in cool, moist environments such as the vegetation along river banks.

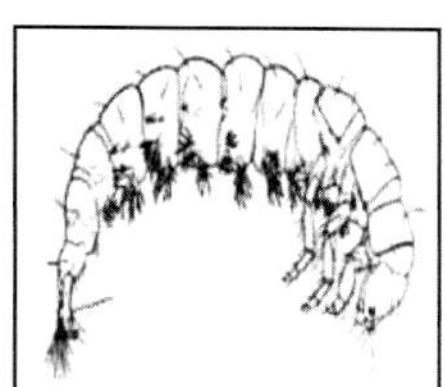

Characteristics features

- This order is closely releted to order lepidoptera.
- Lepidoptera wings cover with scale but tricoptera wing with hairs.
- Larvae of tricoptera resembeles with larvae of lepidoptera (eruciform) caterpillar-like.

Immatures

- Larvae of tricoptera are aquatic and usually enclosed in protective case is covered with stones, silk, leaves, twigs, or other natural materials.
- Mouth part with well develop mandible.
- Antennae short.

- Head capsule well-developed with chewing mouthparts, short antennae
- They have thread-like abdominal gills usually present in case-makers.
- One pair of hooked prolegs often present at tip of abdomen.

Examples of Caddisfly Cases

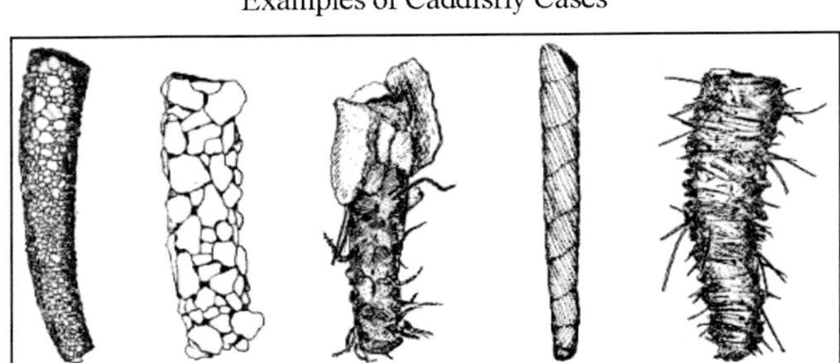

Adults

- The adult of caddisfly are short-lived and survive only 1–2 weeks, but can sometimes last for 2 months.
- Mouthparts reduced or vestigial
- Adult is not feeding stage. they survive for purpose of mating, after mating they die.
- Long antennae filliform and large eyes.
- They have 2 pairs of wings clothed with long hairs, fore wings slightly longer, wings held tent-like over abdomen.
- Female often lays eggs in a gelatinous mass

  Ex. *Chimarra* spp.

7.6. Order Lepidoptera (Lepido-Scales; Pteron - wing)

Ex. Butterflies and Moths

Habitat: Most lepidopteran larvae are herbivores (except some; Parasite- *Epiricannia melanoleuca* (Epipyropidae) - Parasite on Nymph or adult of Pyrilla. Predator- Ape butterfly: Lycaenidae - *Spalgis epius*); some species eat foliage, some burrow into stems or roots, and some are leaf-miners. Although many lepidoptera are valued for their beauty, and a few are useful in commerce (e.g., the silkworm, Bombyx mori), the larvae of these insects are probably more destructive to agricultural crops and forest trees than any other group of insects.

- Lepidoptera (moths and butterflies) is the second largest order in the class insecta.
- Small to large insects with body and wings covered with small, overlapping scales and other appendages giving various beautiful colours to the insects

- Head relatively small free with small neck. Compound eyes are relatively large, two ocelli present one on each side close to the margins of compound eyes.
- Antennal of butterflies: knobbed or hooked at tip while moths: thread-like, spindle-shaped, or comb-like (bipectinate)
- Mouthparts siphoning type form a long coiled tube (proboscis) beneath the head. It is formed by the galea of maxillae.
- Maxillary palpi small or lacking. Mandibles nearly always lacking except in one family micropterygidae with chewing type
- Labial palpi usually well developed.
- Forewings usually large. In males of various insects, groups of more specialized scales known as androconia occur on the upper surface of wings serving as outlets of odoriferous glands. These are fringed distally with each tip finely divided.
- Larvae are called caterpillars usually eruciform. Most of the larval stages are phytophagous and are very serious pests of crops.
- Caterpillars are with well developed head and cylindrical body consisting 13 segments (3 thoracic and 10 adbominal).
- Head bear 2 ocelli on each side and very short bristle like antennae.
- Mouth parts mandibulate with well developed mandibles.
- Labium with a spinneret, a median process for spinning silk.
- Each of the thoracic segments bears a pair of legs which end in a point.
- Abdominal segment 3 to 6 and 10th usually bear a pair of prolegs which are fleshy and broad bearing a number of tiny hooks known as crochets at their end.
- Caterpillars have well developed silk glands and are usually peripneustic

Difference between moth and butterfly

Moth	Butterfly
• Nocturnal	• Diurnal
• Eggs laid in batches	• Eggs laid in singly
• Antennae: plumose,pectinate or filiform	• Clubbed or knobbed
• Pupa usually in cocoon	• Pupa usually in a bare/naked
• Eggs generally flat and round	• Eggs are cigar shaped
• Wing held flat over the body	• Wing vertically upright

Difference sub-order

Zeugloptera	Monotrysia	Ditrysia
• Adult posses functional mandible, maxillae with well developed laccina. • Larvae posses 8 pairs of abdominal legs • Pupa exarate with functional mandible	• Adult do not posses functional mandible, maxillae without lacina. • Larvae posses 7 pairs of abdominal legs	• Adult do not posses functional mandible, maxillae without laccina. • Larvae posses not more than 5 pairs of abdominal legs • Pupa obtect

Butterfly

Family : Nymphalidae

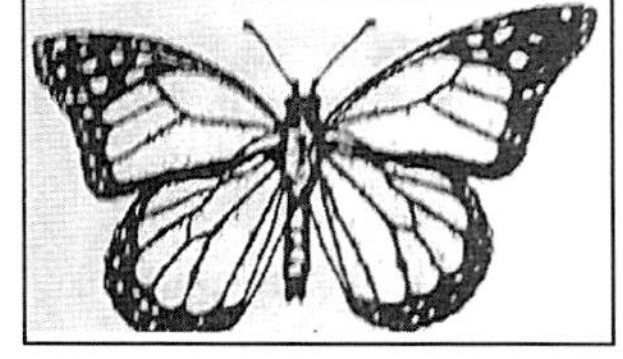

(Brush-footed butterflies, Four-footed butterflies)

- Front legs short, functionless, hairy and folded on thorax
- Larva with many processes or spines on body
- Some species look remarkably like dead leaves, or are much paler, producing a cryptic effect
- Ex. Leaf butterfly, *Kallima* spp.; Castor butterfly, *Ariadne merione* (a defoliator)

Family : Danaidae

(Milkweed butterflies)

- Fore legs short and clawless
- Adults reddish-orange with black and white markings.
- Larvae brightly coloured with fleshy processes on mesothorax and abdomen; feed on various species of milkweed
- Ex. Monarch butterfly, *Danaus plexippus*

Family: Pieridae

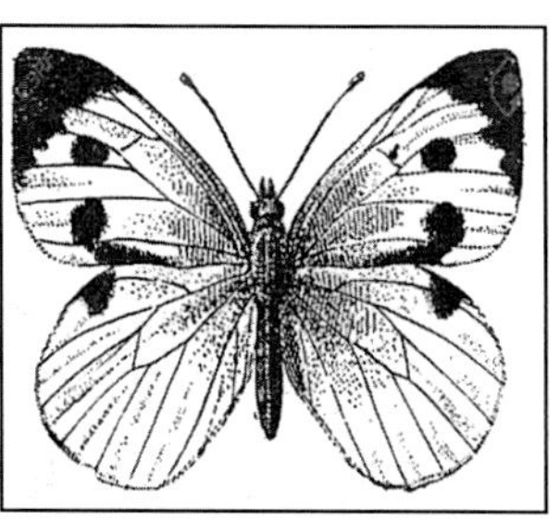

(Whites yellows, Sulphurs)

- Adults white or yellow or orange with black markings
- Front legs well-developed
- Sexes usually differ, often in pattern or number of black markings
- long, often green and usually covered with patches of short, white hairs
- Pupae often elongate chrysalids

- Caterpillars of a few species, such as *Pieris brassicae* and *Pieris rapae* are notorious pests

Family: Papilionidae

(Swallow-tails)

- Large, brightly coloured
- Front legs well-developed
- Characteristic tail-like piece extension on hind wings
- Only one anal vein in the hind wing
- Larval smooth with with a series of fleshy dorsal tubercles
- Ex. Swallowtail, *Papilio glaucus;* Lemon butterfly, *Papilio demoleus*

Family: Lycaenidae

(Blues, coppers, Hairstreaks)

- Brightly coloured, upper wing surface either metallic blue or coppery; lower surface light coloured. Colouration differ in sexes
- Eyes surrounded by white rim and antennae have alternating black and white bands
- Hind wings often with delicate tail like prolongations and 2 – 3 black spots
- Larvae often flat, with glands that secrete ants attracting substance
- Phyto- or entomophagous feeding on aphids, scale insects, and ant larvae. Most species live in association with ants, a relationship called myrmecophily. These associations can be mutualistic, parasitic, or predatory
- Ex. Pomegranate fruit borer, *Virachola isocrates*

Family: Hesperiidae

(Skippers)

- Body moth-like, small wings
- Antennal club hooked at tip
- Larval head big, attached to a neck-like collar
- Larvae spins cocoons
- Ex. Rice skipper, *Pelopidas mathias*

Moths

Family: Tineidae

(Clothes moths, Fungus moths)

- Head with rough hairs, long maxillary palpi
- Wings held roof-wise over body when at rest

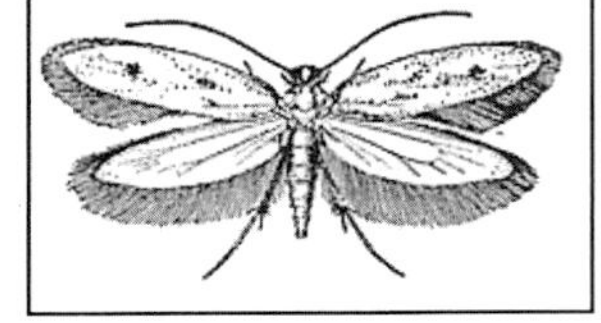

- Most larvae feed on fungus, lichens, and detritus; some phytophagous
- Some larvae construct cases and feed on natural fibers
- Ex. *Tinea* spp. Pests include webbing clothes moth and case making clothes moth which damage stored fabrics

Family: Gelechiidae

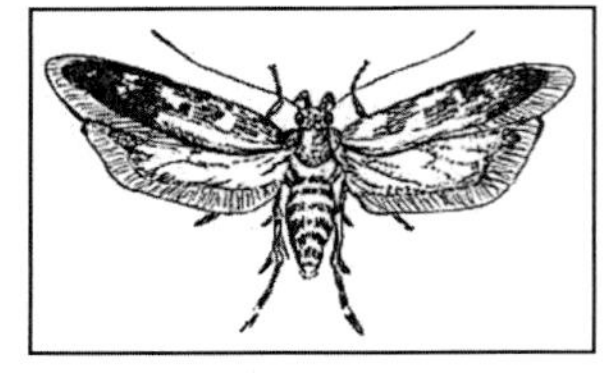

(Twirler moths, Gelechiid moths)

- Small moths with narrow, fringed wings
- Forewings trapezoidal and narrower than hindwings
- Larvae feed internally on plants or plant products, and bore into seeds, tubers, and leaves; sometimes causing galls

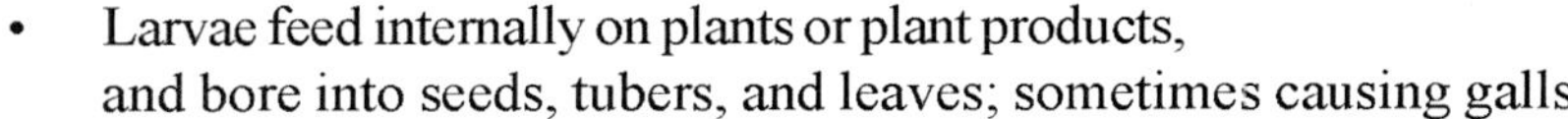

- Larvae of some species are suitable for biological control of invasive plants, weeds
- Ex. Angoumois grain moth, *Sitotroga cerealella;* pink bollworm, *Pectinophora gossypiella*

Family: Pyralidae

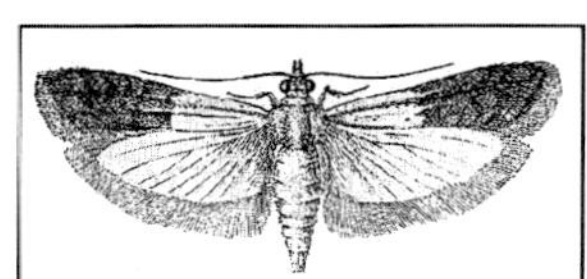

(Pyralid moths, Snout moths)

- Small moths, with beak-like proboscis covered with scales
- Legs long
- Fore wings triangular, hind wings broader
- Larvae with varied habits, feed on a wide variety of plants and plant products including stored grain, dried fruits and tobacco
- Pests include: European corn borer (*Ostrinia nubilalis*), Indian meal moth (*Plodia interpunctella*), and the greater wax moth (*Galleria mellonella*)

Family: Lasiocampidae

(Tent caterpillar, Eggars, Lappet moths)

- Protruding mouthparts of some species that resemble a large nose
- Decorative skin flaps found on caterpillar's prologs
- Caterpillars large, often hairy, especially on their sides
- Feed on leaves of trees and shrubs, and often use these same plants to camouflage their cocoons. Some species are called tent caterpillars due to their habit of living together in nests spun of silk
- Ex. Tent caterpillar *Malacosoma* spp.

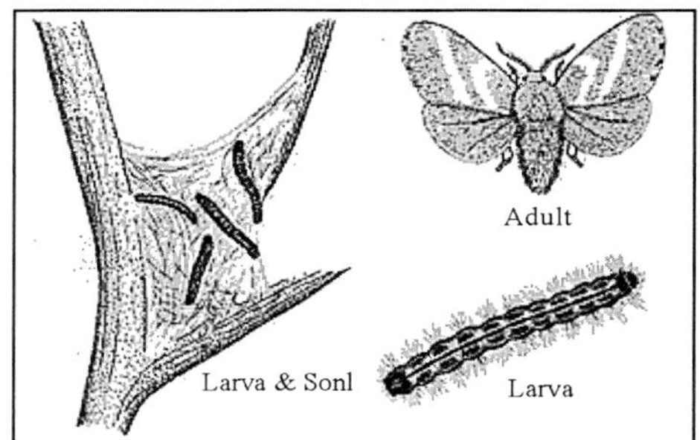

Family: Saturniidae

(Moon moths, Giant silk moths)

- Large colourful moths with thick, hairy bodies
- Reduced mouthparts, and small head
- Antennae bipectinate or feathery;
- Larvae covered in tubercles and spines, while pupae are wrapped in silk cocoons that are attached to twigs and leaves
- Larvae feed on a wide range of trees and shrubs; a few species are important defoliator pests
- Ex. Tussor silkworm, *Antheraea paphia*

Family: Sphingidae

(Hawk moths, Sphinx moths, Horn worms, Hummingbird moths)

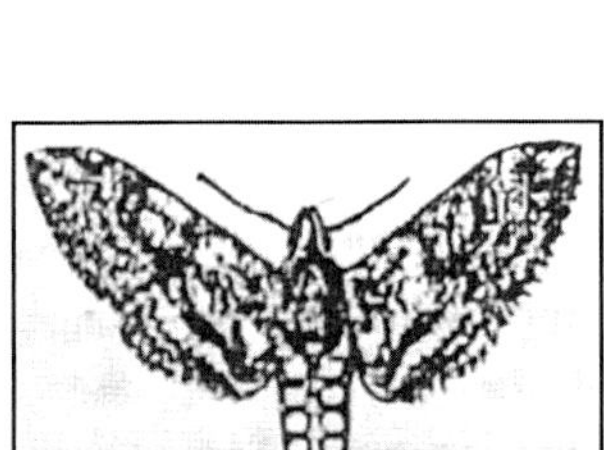
Hawk moth

- Medium to large adults; powerful fliers; with long proboscis for collecting nectar
- Body thick, hairy
- Antennae thickest near midpoint, sometimes bipectinate
- Long, narrow forewings, hind wings shorter
- Larva smooth with a dorsal horn (anal horn) on the 8th abdominal segment
- Ex. Tobacco hornworm, *Manduca sexta;* Death's head moth, *Acherontia styx*

Family: Noctuidae

(Loopers, Owlet moths, and Underwings)

Cut warm

- Labial palpi well developed; antennae filiform
- Front wings mottled brown in colour, hind wings differ in colour and/or pattern
- Many species are nocturnal and have a tympanal auditory organ on base of abdomen
- Some larvae are semiloopers, with either 3 or 4 pairs of prologs

- Most larvae feed at night on leaves and also stem borers; many species are agricultural or horticultural pests
- Ex. Fall armyworm, *Spodoptera frugiperda*; black cutworm, *Agrotis ipsilon;* cabbage looper, *Trichoplusia ni*; gram pod borer, *Helicoverpa armigera;* tobocco cut worm, *Spodoptera litura*

Family: Arctiidae

(Tiger moths)

- Antennae filiform
- Wings colorfully marked with dark spots or bands
- Nocturnal and attracted to light
- Larva either sparsely hairy or densely hairy (wooly bear)
- Ex. Red hairy caterpillar, *Amsacta albistriga*; Bihar hairy caterpillar, *Spilosoma obliqua*

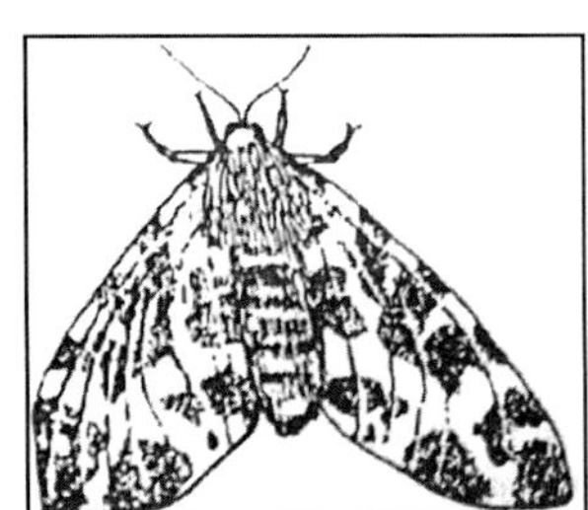

Tiger moth

Family: Plutellidae

(Diamondback moths)

- Small to medium sized
- Head usually with smooth scales and antennae often thickened in middle
- Wings elongated and hind wings often bear long fringes; fore wings sickle-shaped
- Adults mostly nocturnal or crepuscular
- Larvae feed on leaf surface
- Ex. Diamondback moth, *Plutella xylostella*

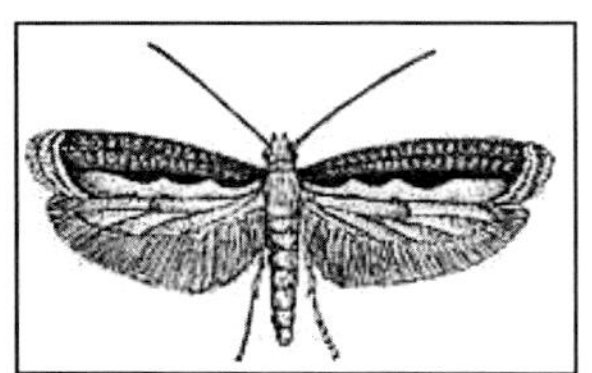

Family: Geometridae

(Loopers, Geometer moths)

- Male antennae often feathered Wings angular, thin, females have reduced wings in some species
- With distinctive paired tympanal organs at base of abdomen (lacking in flightless females)
- Larvae move in a looping fashion and called loopers, spanworms, or inchworms
- Several species are notorious pests
- Many species are cryptic, caterpillar camouflaged as a broken twig

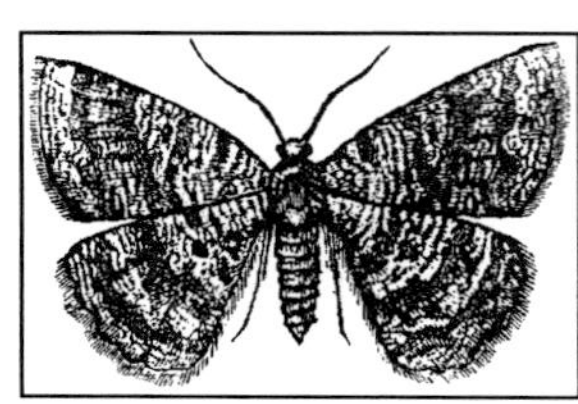

- Larvae with only 2 or 3 pairs of prologs; plant feeders
- Ex. Tea looper, *Biston suppressaria.*

Family: Gracillariidae

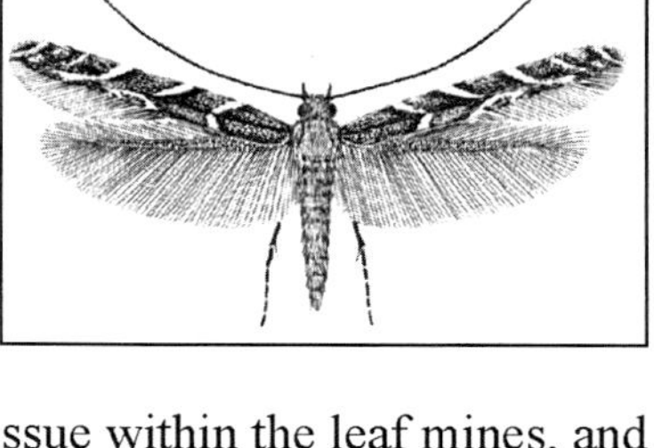

(Leaf miners)

- Small moths, with narrow fringed wings
- Larvae apodous, 1st to 5th instars flat, with specialised mouthparts adapted for sap feeding. Older instar larvae have normal chewing mouthparts for feeding on plant tissue within the leaf mines, and have a fully functional silk-producing organ, the "spinneret".

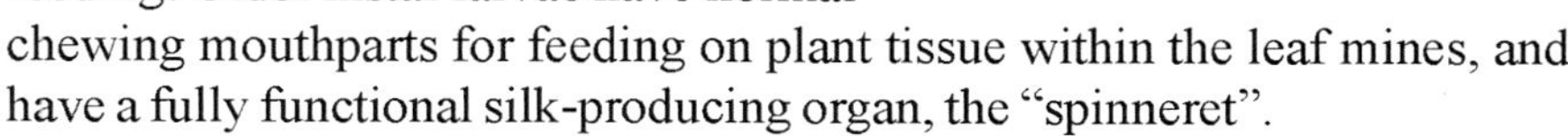

- Mostly leaf-miners, includes several economic pests and invasive pest species
- Ex. Litchi fruit and shoot borer, *Conopomorpha cramerella*

Family: Bombycidae

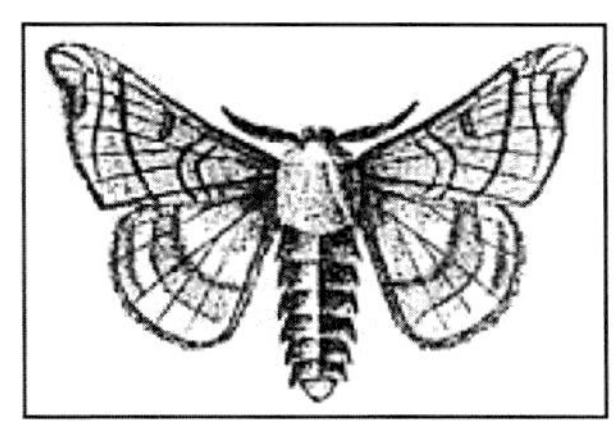

(Silk worm moths)

- Antennae bipectinate
- Larva smooth with prologs
- A prominent horn in centre of 8th abdominal segment
- Pupation occurs in dense silken cocoon
- Ex. Silkworm, *Bombyx mori* (basis of silk industry)

Family: Psychidae

(Bagworm moths, Bagmoths)

- Female apterous in many species, males winged
- Antennae bipectinate
- Caterpillars construct cases of silk and materials like sand, plant materials, etc. which are attached to rocks, trees, fences during pupal stage, but are otherwise mobile
- Caterpillars feed on lichens, leaves
- Ex. Bagworm, *Clania crameri*

Family: Crambidae

(Crambine snout moths, Grass moths)

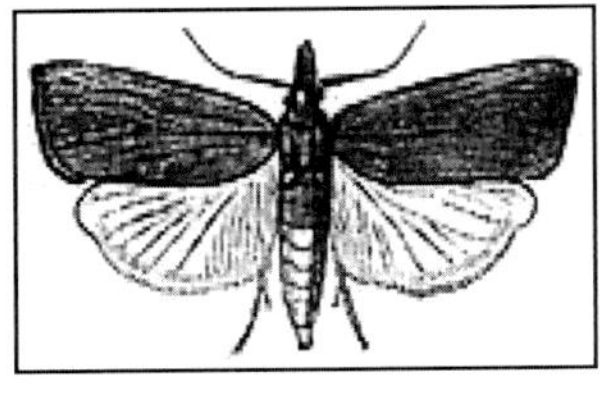

- Small moths, proboscis with scales at base
- Fore wings narrow and long
- Tympanal organ at base of abdomen
- Larvae remain in silken webs or galleries
- Many species are serious pests
- Ex. Spotted stalk borer, *Chilo partellus*; Striped rice stem borer, *Chilo suppressalis*; Sugarcane shoot borer, *Chilo infuscatellus*; Bean pod borer, *Maruca vitrata*

7.7. Order: Siphonaptera (Siphon-tube like;aptera-wingles)

Ex. Fleas

Habitat: As adults, all fleas are blood-sucking external parasites. Most species feed on mammals, although a few (less than 10%) live on birds. Only adult fleas inhabit the host's body and feed on its blood.

Characteristics

- They are popular known by spread of plague disease.
- Adults depend for food on blood of mammals and birds.
- Fleas are ectoparsites which live freely while larvae are detritivores.
- Active with a hard exoskeleton, strong hind legs with powerful leg muscles for jumping.
- In case of flea the body laterally compressed and usually has caudally directed setae and spines.
- Without eyes, 2 simple ocelli may be present
- Antennae short and stout
- Mouthparts piercing for sucking blood
- Both adults and larvae have many backward pointing hairs
- Secondarily wingless
- Larvae worm-like (vermiform), with thin covering of bristles

  Ex. Oriental rat flea, *Xenopsylla cheopis,*

  - Primary vector of bacterial pathogen for bubonic plague

7.8. Order: Strepsiptera (Twisted-wing parasites)

Ex. Stylopids, Stylops, Strepsipterans

- Females larviform, legless and wingless, mostly eyeless, look like grub, partially projecting from host's abdomen
- Reduced mandibulate mouthparts
- Adult male with small head and protruding compound eyes
- Antennae multi-segmented, branched
- Fore wings reduced to small, club-like structures; hind wings very large and fan-shaped
- Males emerge with adult-like body
- Larvae hypermetamorphic; 1st instar (triungulin) with legs; high mobility
- Mostly live as internal parasites of bees, wasps, grasshoppers, leafhoppers, and other Hemiptera

  Ex. *Stylops* sp.

7.9. Order: Mecoptera (Meco - long and ptera- wings)

Ex. Scorpionflies, Hangingflies

Habitat: The Mecoptera (scorpionflies) are a curious group of terrestrial insects that usually live in moist sylvan habitats. Both larvae and adults are omnivorous. Mostly, they feed upon decaying vegetation and dead (or dying) insects. Larvae generally remain in the soil; they have chewing mouthparts and resemble caterpillars (Lepidoptera) or white grubs (Coleoptera).

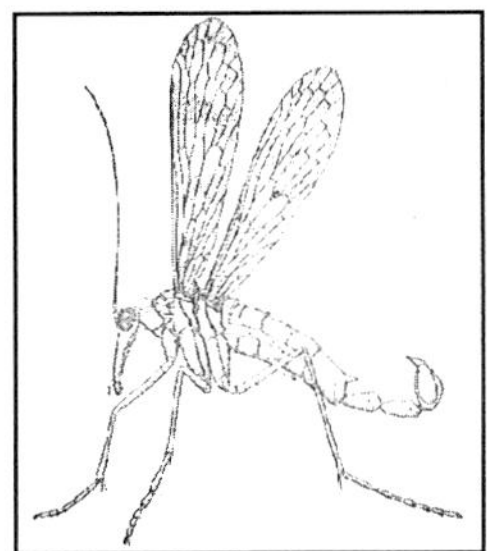

Scorpionfly

- Head elongate; mouthparts mandibulate
- Eyes large
- Antennae long, multi-segmented
- Wings narrow, elongate, and similar; many crossveins, and often mottled with patches of colour. Some species are secondarily wingless or with reduced wings
- Males of some species have enlarged external genitalia held recurved over abdomen like a scorpion's tail
- Larvae remain in soil, body eruciform (caterpillar-like) or scarabaeiform (grub-like); with 8 pairs of prologs
- Ex. *Panorpa germanica*

7.10. Order: Diptera (Di –Twice; pteron- wings)

Ex.True flies, Mosquitoes, Gnats, Midges

Habitat: The Diptera probably have a greater economic impact on humans than any other group of insects. Some flies are pests of agricultural plants; others transmit diseases to humans and domestic animals. On the other hand, many flies are beneficial particularly those that pollinate flowering plants, assist in the decomposition of organic matter, or serve as biocontrol agents of insect pests.

Characters feature

- Small and soft bodied insects with prominent head and small neck.
- Eyes larger in males compare to female.
- Ptilinum or frontal sac is important character of cyclorrhapha indicated by the frontal or ptilinal suture.
- It is a retractile bladder like organ employed to break open the puparium
- Antennae filiform, stylate, or aristate and mostly 3 segmented (except in Nematocera).
- Mouthparts suctorial (haustellate) type usually forming a proboscis.
- In many they are piercing and sucking and in others they are sponging (lapping) with labium distally expanded in to a pair of fleshy lobes.
- Mesothorax larger than pro- or metathorax and supporting the functional wings,
- One pair of wings (front) and hind wings reduced (halteres) which act as balancers
- Legs well developed,
- Tarsus 5 segmented pulvilli and an empodium usually present
- Holometabola i.e. complete metamorphosis (egg, larva, pupa, adult)
- Larvae eruciform and apodous known as maggots mostly amphipneustic
- Pupa either free or enclosed in the hardened larval cuticle known as puparium (coarctate pupa)

The Diptera have traditionally been divided into three suborders:

1. Nematocera (flies with multisegmented antennae)
2. Brachycera (flies with stylate antennae)
3. Cyclorrhapha (flies with aristate antennae)

In some newer classifications, Brachycera includes the Cyclorrhapha.

S.No.	Character	A. Cyclorrhapha	B. Nematocera	C. Braachycera
1.	Larvae act	Vertically	Horizontally	Vertically
2.	Antenna type	Aristate	Antenna long many segmented	Antenna short and few segmented
3.	Pupa	Coarctate	Obtect	Exarate
4.	Emergence of adult	Adult emerge through pressure applied by an eversible bladder **Ptillinum** in the head.	Adult emerge through a straight spilt in the thoracic region.	Adult emerge through a straight spilt in the thoracic region.
5.	Example	House fly, hower fly, pamace flies.	Mosquitoes and midges	Robber flies and Horse flies.

Sub order: Nematocera

Family Culicidae

Ex.Mosquitoes

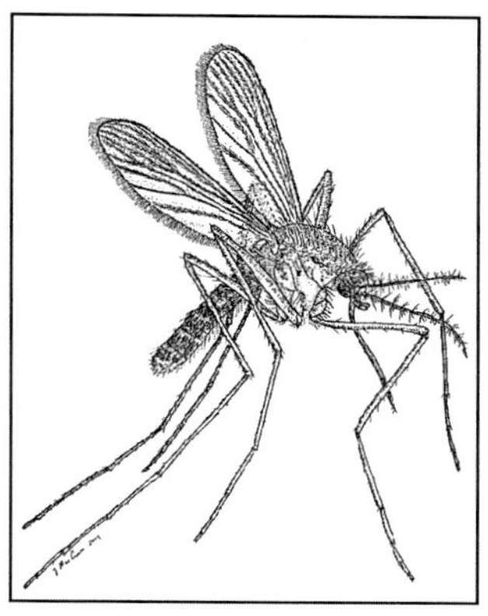

- Piercing-sucking type mouth with proboscis
- Long maxillary palps and highly plumose antennae in male while pilose in female have short maxillary palps with less plumose antennae.
- Hump-backed thorax, legs delicate
- Wings fringed and scaly
- Females blood suckers, efficient disease vectors may spread malaria, encephalitis, yellow fever, filariasis, and other diseases
- Larvae and pupae aquatic, develop in shallow stagnant waters
- Larvae are called wrigglers and pupa known as tumbler

  Ex. Common house mosquito, *Culex pipiens* fatigans

  Yellow fever mosquito, *Aedes aegypti*;

Family : Cecidomyiidae

Ex. Gall midges

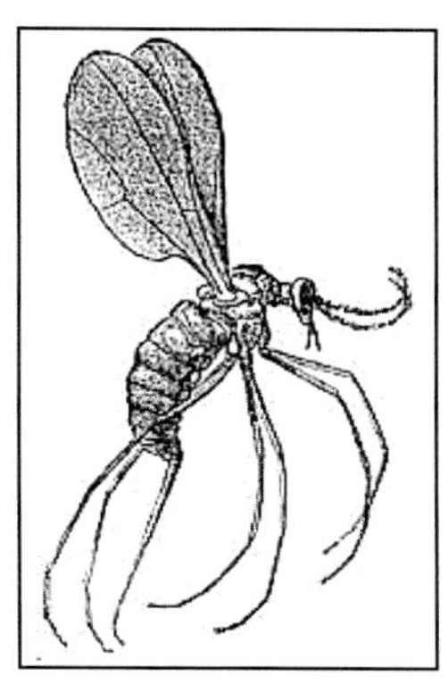

- Minute delicate, mosquito-like flies with long legs and monoliform antennae.
- Mouthparts reduced.
- Tibiae without spur
- Wing venation reduced wings hairy.
- Larvae form plant galls.
- Some species are also scavengers, predators, or parasites

Ex. Rice gall midge, *Orseolia oryzae (*maggot's feeding produces galls)
Hessian fly, *Mayetolia destructor*

Family: Tipulidae

Ex. Daddy-long legs, Crane flies

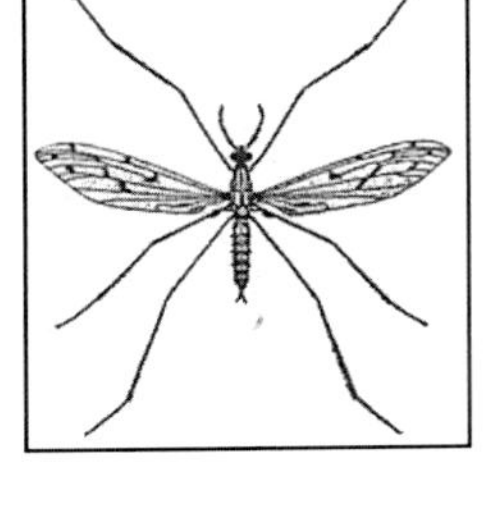

- These insects have very long slender legs and halteres conspicuous.
- Antennae are multisegmented pectinate or serrate type.
- V-shaped suture on dorsal side of mesothorax.
- Wing with 2 anal veins.
- Abundant in damp areas or where there is lush vegetation.
- Larvae live in soil, mud, mosses, in dying and decaying wood. Feed mainly on plant remains and detritus.

Ex. *Tipula* spp.

Family: Chironomidae

Ex. Gnats, Midges

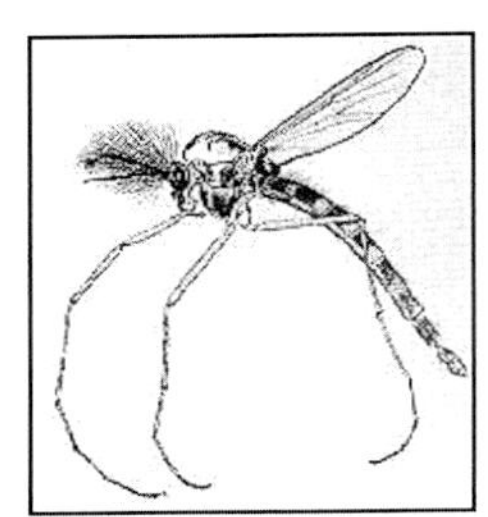

- Antennae plumose in male and pilose (hairy) in female.
- Chewing mouthparts with short palps.
- Long slender legs.
- Larvae found in decaying matter, under bark, or burrowing within moist ground
- Aquatic larvae of some species have haemoglobin in blood and red in colour (called blood-midges).
- They usually live in the mud and feed on organic matter.

Ex. *Chironomus* spp.

Family: Psychodidae

Ex. Moth flies, Sand flies

- Small to medium sized, moth like, usually densely pubescent.
- Ocelli absent.
- Wing hairy.
- Larvae occur in a wide range of habitats
- Mainly nocturnal and during day rest in dark places. Feed on human blood and important disease vectors.
- They spread leishmaniasis (*kala-azar*), sand fly fever, and other diseases

Ex. *Phlebotomus* spp.

Sub order: Brachycera

Family: Tabanidae

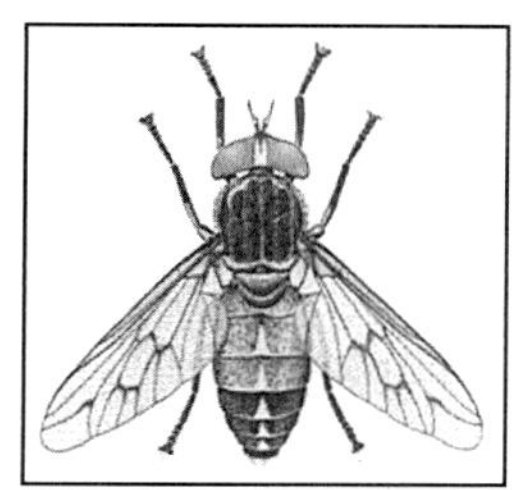

Ex. Horse flies, Deer flies

- Large bright coloured eyes cover most of head, devoid of bristles.
- 3rd antennal segment elongate and annulated
- Females with powerful, often long piercing proboscis.
- Wing clear, or spotted; swift fliers .
- May spread trypanosomiasis, and other diseases
- Male feeds on nectar; female sucks blood from cattle and horses.
- Larvae aquatic, semi-aquatic or terrestrial; predators but also able to feed as facultative saprophages

  Ex. *Tabanus* spp.

Family: Asilidae

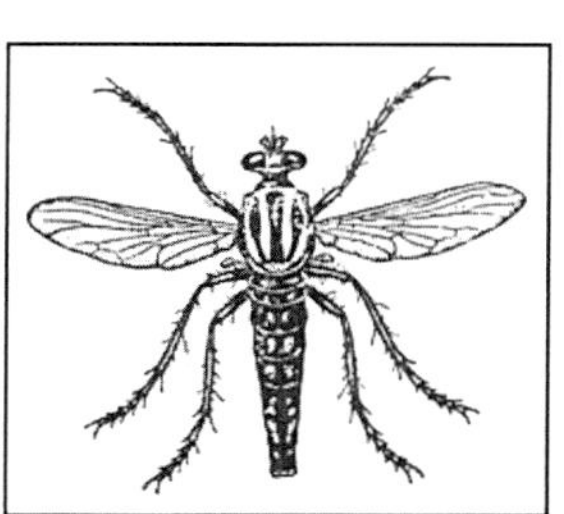

Ex. Robber flies, Assassin flies

- Stylate powerful mouthparts surrounded by long hairs
- Concave depression on top of head between eyes.
- Legs stout, hairy and suited for catching prey
- Larvae live in soil and decaying wood and feed on larvae of other insects
- Important predators and act as biological control agents of many pests

  Ex. *Asilus* spp.

Suborder: Cyclorrhapha

Family: Muscidae

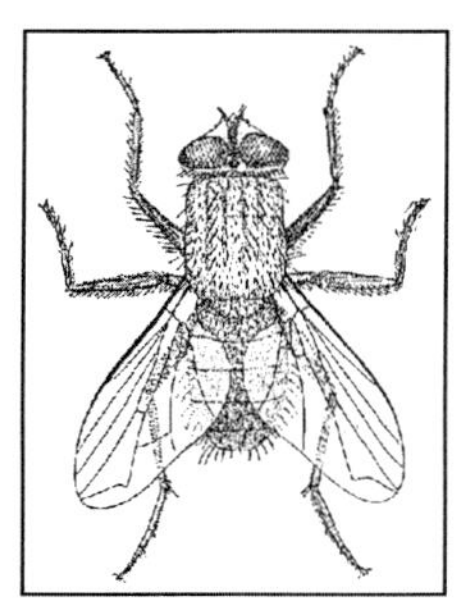

Ex. House fly

- Arista bare to plumose, mouthparts sponging .
- Pretarsus consists of 2 claws and two adhesive pads
- Maggots scavengers or carnivorous.
- Adults carry certain disease (Cyclorrhapha) such as as dysentery, cholera, and yaws may be transmitted on their feet, body hairs and mouthparts.
- These are among the most cosmopolitan of all insects.
- Some species have biting mouthparts, others are merely scavengers.

  Ex. Common house fly, *Musca domestica*

Family: Tephritidae

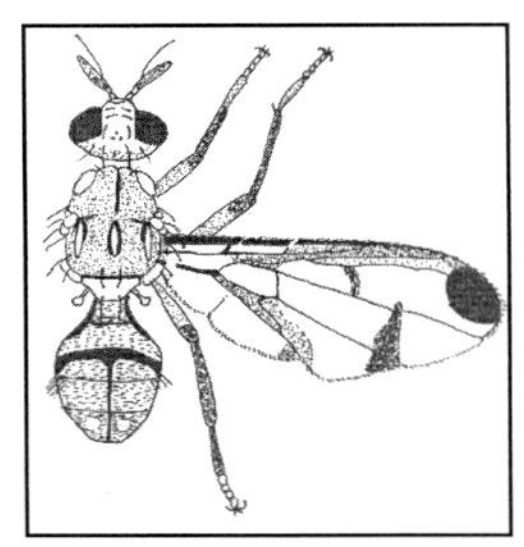

Ex. Fruit flies

- Subcosta bends apically and fades out
- Wings spotted or banded
- Female with a sharp and projecting ovipositor
- Larvae phytophagous, many species are agricultural pests and cause direct damage to fruits and vegetables.

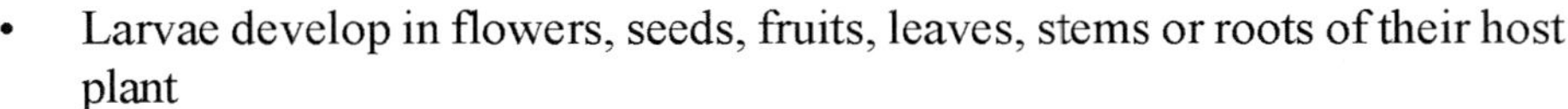

- Larvae develop in flowers, seeds, fruits, leaves, stems or roots of their host plant

Ex. Cucurbit fruit fly, *Bactrocera cucurbitae*

Oriental fruit fly, *Bactrocera dorsalis*

Family: Drosophilidae

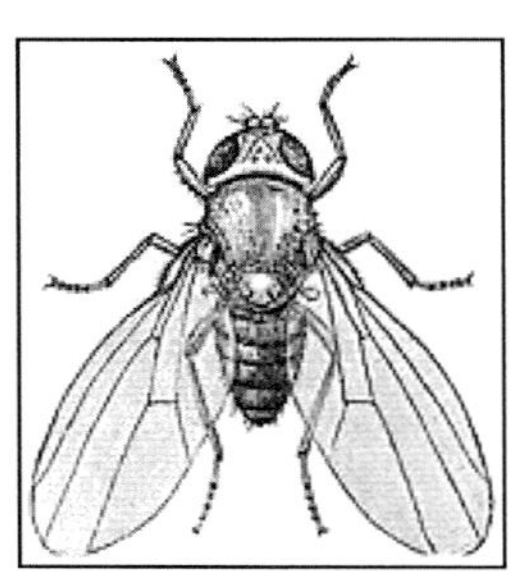

Ex. Vinegar gnats, Pomace flies

- Minute to medium sized, yellow flies
- Eyes usually red; ocelli present
- Arista usually with long rays, in some cases pubescent
- Wing usually unmarked
- Attracted to rotting vegetables and fruits, fungi
- Larvae feed on yeast and products of fermentation
- Extensively used in the study of animal genetics

Ex. *Drosophila melanogaster*.

Family : Calliphoridae (Blow flies)

- Metallic body color
- Head, body and legs often with strong bristles
- Arista plumose
- Calypters (alula- flap at wing base) large
- Scavenger, lay eggs on dead animals. The larvae feed on decaying tissue of animals. They are efficient disease vectors of dysentery
- Some species important in forensic entomology.
- Larvae can be true obligate agents of myiasis in vertebrates

Ex. Screwworm, *Cochliomyia hominivorax*; Common blue bottle, *Calliphora erythrocephala*

Family : Syrphidae (Flower flies, Hover flies)

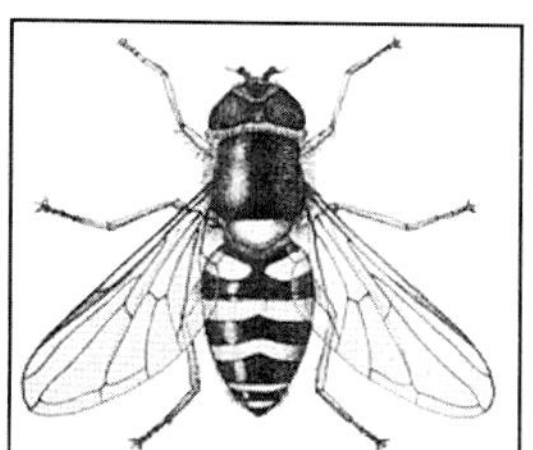

- Birghtly coloured and brilliantly striped flies that mimics bees, bumblebees, or wasps.
- Head without bristles.
- Many larvae are predaceous on aphids or live in decaying vegetation; also feed on organic detritus, plant materials.
- Abdomen with distinct black and yellow markings
- Maggots prey on soft bodied insects especially aphids
- Adults hover over flowers, feed on pollen and nectar and help in pollination

  Ex. *Syrphus* spp.

Family : Agromyzidae (Leaf miners)

Ex. Small, black or yellow flies

- Wing with costal break present at the apex of Sc
- Female anterior part of abdominal segment 7 forming an oviscape
- Larvae feed in living plant tissue. About 3/4 species are leaf miners, while remainder feed in stems or form galls; several species are agricultural pests

  Ex. *Agromyza* spp. pea stem fly, *Ophiomyia phaseoli*

Family : Hippoboscidae (Louse flies or keds)

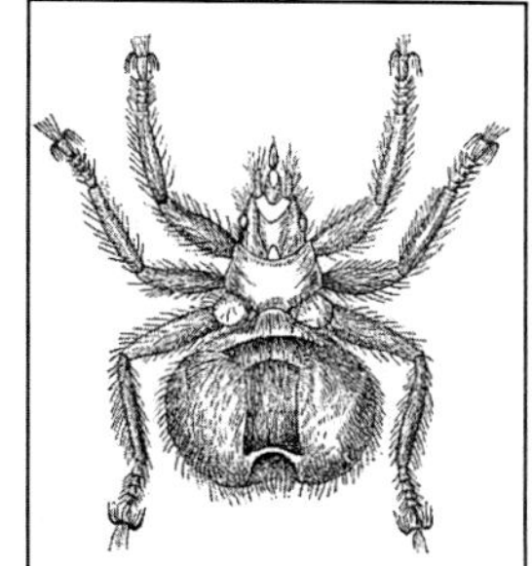

- Body flat, leathery; legs short, strong
- Tarsi with hook-like claws, enabling to cling to the hairs or feathers of their hosts
- Wings present or absent
- Viviparous, give birth to mature larvae which are glued to host's hairs
- Blood sucking ectoparasites on birds and mammals

  Ex. *Hippobosca rufipes* (an ectoparasite of cattle)

7.11. Order: Hymenoptera (Hymen- Membranous; pteron -wing)

Ex. Ants, wasps, Bees, Sawflies, Horntails

Habitat: They are primarily phytophagous ex.Symphyta, the parasitic or gall-making ex. Apocrita, the non-social Apocrita whose larvae are fed prey or pollen captured by the mother, and the social Apocrita. Virtually all apocritan larvae are provided with a food resource and do not forage. Hymenoptera are generally diurnal or crepuscular, with few nocturnal species. Adults are often found among vegetation, particularly flowers, or among leaf litter (e.g. ants).

Many build nests or burrows in the ground. Social species usually create large nests that may be well defended by stinging worker. It is the most important order comprising of parasites, predators and bees involved in pollination and honey production and this order most of insects are social living.

Characteristic features

- Insect having Small to medium size and shape.
- Head extremely mobile.
- Compound eyes well developed and ocelli usually 3 or absent.
- Antennae longer in males
- Mouth parts are biting and lapping or sucking type
- Usually two pairs of naked membranous wings are present with reduced venation.
- Hind wings are smaller and have a row of tiny hooks (hamuli) on their anterior margin by which they attach to the front wings.
- Stigma is present in the forewings along the costal margin near the apex.
- Legs slender, trochanter 1 or 2 segmented (called "trochantellus")
- Abdomen basally constricted to form pedicel or petiole.
- The 1st abdominal segment fused with metathorax called as propodaeum.
- Second segment forms pedicel.
- The remaining region of the abdomen is known as gaster.
- Ovipositor modified for sawing, boring, piercing, stinging etc.
- Larvae are called as grubs and generally apodous or may be polypod.
- Pupa adecticous and exarate sometimes cocoon is also found.
- Metamorphosis complete and complex.

Classification of order

S.No.	Sub order: Symphata	Sub order: Apocrita
1.	Abdomen broadly attached with thorax and no constriction between 1st and 2nd abdominal segment.	Abdomen deeply constricted or petiolated between 1st and 2nd abdominal segment.
2.	Legs with 2 segmented trochanter	Legs with 1 or 2 segmented trochanter
3.	Prologs without crochets	Prologs absent
4.	Ovipositor is saw like and suited for piercing the plant tissue.	Ovipositor is not saw like and is suited for piercing or stinging.
5.	-	Fore tibia with one spur
6.	Larva is caterpillar and eruciform type.	Larva is grub type and is apodous, eucephalous type.
7.	Stemmata are present.	Stemmata are absent.
8.	Behavioural sophistication is less.	Behavioural sophistication is more.
9.	Phytophagous	Generally parasitic.

Sub order: Symphyta

Family : Tenthredinidae

Ex. Sawflies

- Black or brown
- Lack slender "wasp-waist", or petiole, between thorax and abdomen
- Females use their saw-like ovipositors to cut slits through barks of twigs, into which eggs are laid, which damages trees
- Larvae are typically herbivores and feed on the foliage of trees and shrubs, some leaf miners, stem borers, or gall makers.

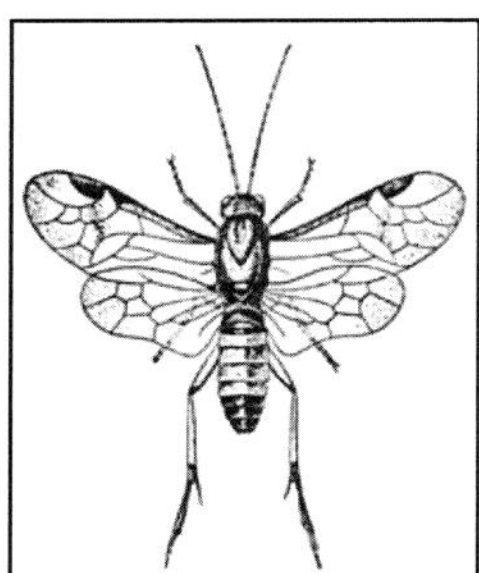

- Larva is caterpillar belonging to eruciform type.
- They are phytophagous.
- Stemmata are present.
- 6-8 pair of prolegs is present.
- In prolegs, crochets are absent.
- In adults, abdomen broadly joins to the thorax.
- Adult saw flies have a pair of cenchri in metanotum.
- Foretibia with 2 apical spurs.

Ex. Mustard saw fly, *Athalia lugens proxima*

Sub order: Apocrita

Family : Ichneumonidae

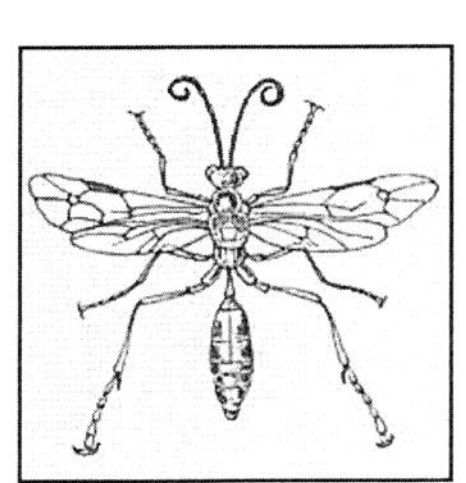

Ex. Ichneumon Wasps

- Long, slender wasps with antennae 16 or more segmented
- Trochanters 2-segmented
- Gaster long, about 3x as long as head and thorax
- Females with a long, slender ovipositor
- Many species are internal parasitoids of immature stages of the host

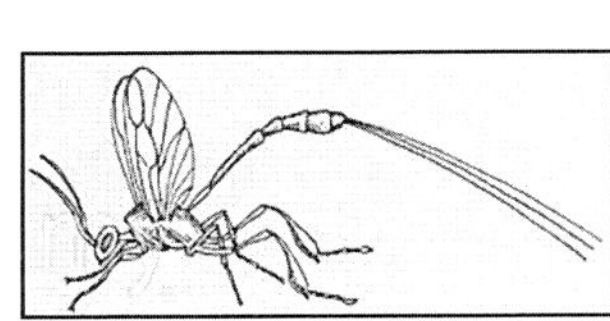

Ex. *Ichneumon spp.*

Family : Braconidae

Ex. Braconid wasps

- Large family of parasitoid wasps
- Often black-brown (sometimes with reddish markings), though some species exhibit striking colouration and patterns

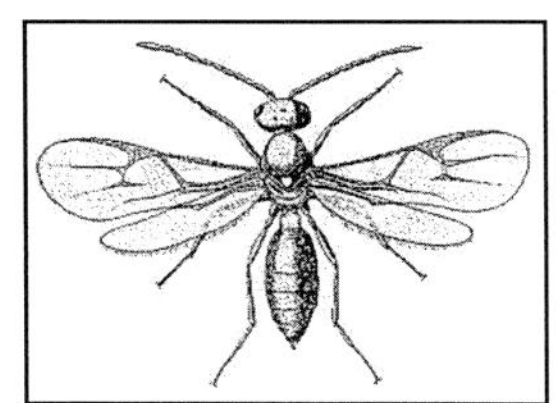

- Antennae 16 segmented or more
- Hind trochanter with 2 segments
- Ovipositor long
- Most internal and external primary parasitoids on other insects; they kill their hosts

  Ex. *Bracon* spp.; *Apanteles* spp.

Family : Encyrtidae

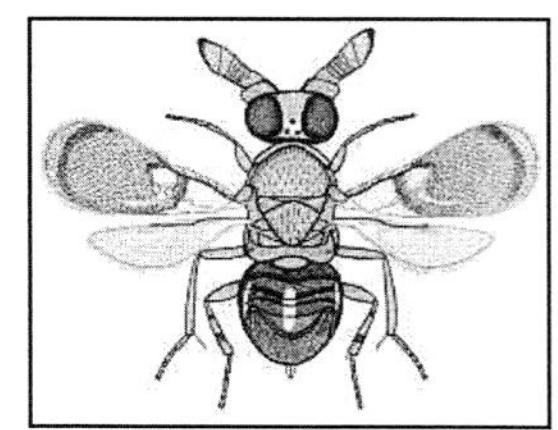

- Small, 1 -2 mm long, found in virtually all habitats
- Mesopleuron broad, convex
- Parasitic wasps, important as biological control agents
- Some species exhibit polyembryony
- Ex. *Encyrtus* spp.

Family : Eulophidae

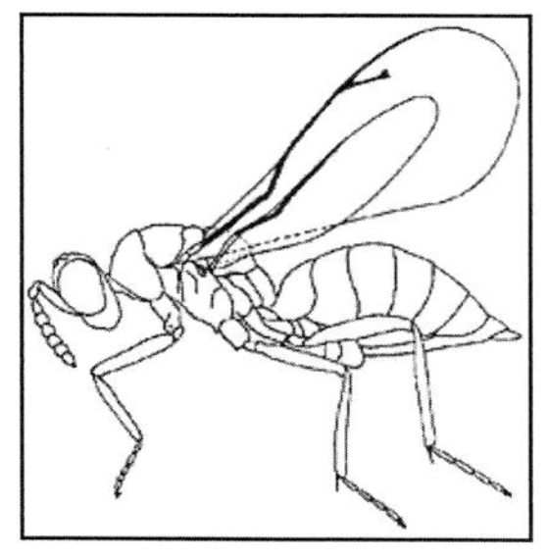

- Small, 1 – 3 mm long,
- 4 tarsomeres on each leg
- Larvae of a very few species feed on plants, but majority are primary parasitoids on a huge range of arthropods
- Some species are hyperparasite
- Found in all habitats (even aquatic)

  Ex. *Elasmus* spp.; *Eulophus* spp.; *Tetrastichus* spp.

Family : Trichogrammatidae (egg parasitoids)

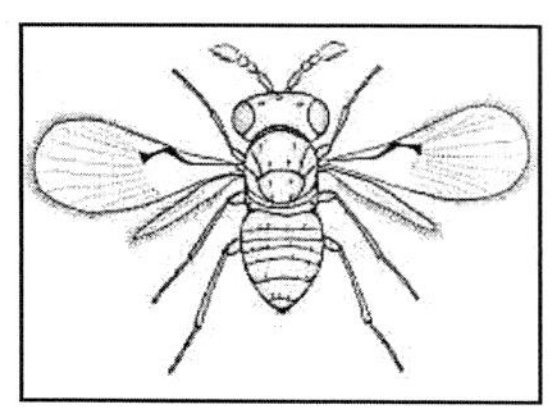

- Tiny wasps (0.3 – 1.0 mm long)
- Parasitize eggs of many different orders of insects
- Important biological control agents, attacking eggs of many pest insects (esp. Lepidoptera)
- Not strong fliers. Fore wings typically somewhat stubby and paddle-shaped, with a long fringe of hinged setae around outer margin
- Males of some species are wingless, and mate with their sisters inside host egg in which they are born

  Ex. *Trichogramma* spp.

Family : Chalcididae (Chalcid wasps)

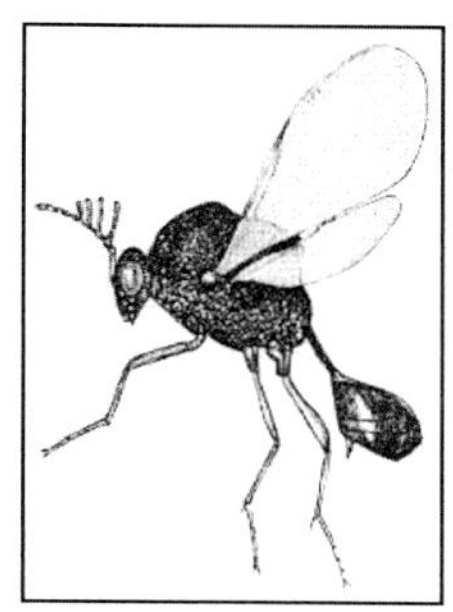

- Often black with yellow, red, or white markings, rarely brilliantly metallic
- Hind femora often greatly enlarged, with a row of teeth or serrations along lower margin
- Ovipositor straight, short
- Mostly parasitoids of different insects and a few hyperparasitoids

  Ex. *Brachymeria* spp.; *Megachalcis* spp.

Family : Vespidae

Ex. Yellow jackets, Hornets, Potter wasps

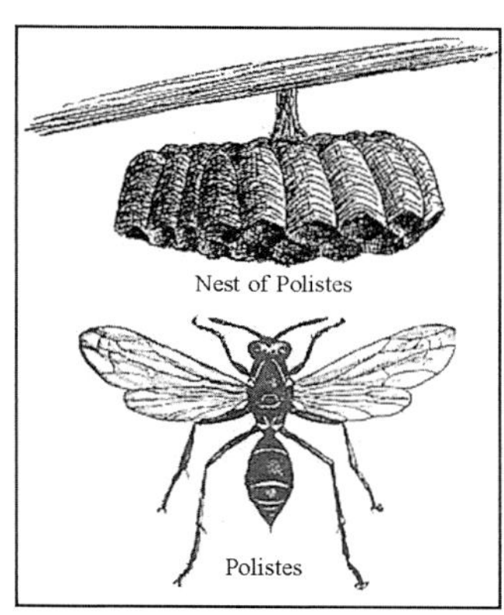
Nest of Polistes
Polistes

- Yellow or red with black markings, with long petiole
- Forewings fold in half longitudinally
- Solitary or as social wasps
- Social Wasps, make paper-like nests tended by sterile female workers
- Predators of caterpillars, collect and store caterpillars in nests
- Many species live in colony comprised of 3 castes, queens, female workers, and drones. Only workers can sting

  Ex. Oriental hornet, *Vespa orientalis*

Family : Formicidae

Ex. Ants

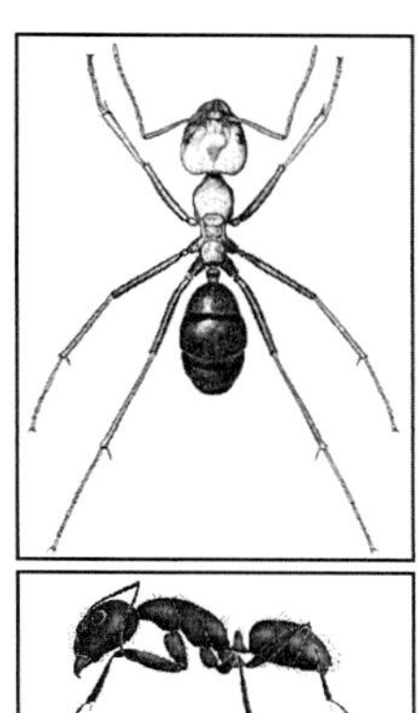

- Antennae elbowed
- Wingless workers; winged swarmers (reproductives - both males and females)
- Peduncle located between thorax and abdomen - first (or first and second) abdominal segments separate from rest of abdomen
- True social insects. Workers (sterile females) forage for provisions (vegetation, seeds, or other insects)
- Found in many different types of habitats, live in nests in various ways
- Colonies of ants vary in size, 3 castes - workers, queen and males

  Ex. Black garden ant, *Lasius niger*; weaver ant, *Oecophylla smaragdina*

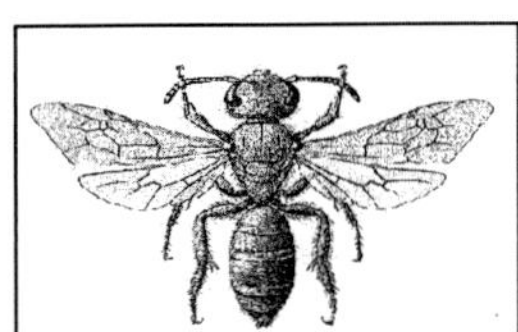

Family : Halictidae

Ex. Mining bees, Sweat bees

- Small sized, body often metallic in color
- Basal vein in fore wing strongly curved
- Sting only if disturbed; sting is minor
- Most halictids nest in ground, in wood, and provide pollen and nectar to larvae
- Many species important pollinators
- Solitary; many species are eusocial at least in part, with fairly well-defined queen and worker castes

Ex. *Halictus* spp.

Family : Apidae

Ex. bumble bees, carpenter bees, orchid bees, cuckoo bees, honey bees

- Body heavily clothed with branched hairs
- Mouthparts form a tongue for collecting nectar
- 1st segment of hind tarsi (Basitarsus) enlarged; usually bearing a "pollen basket"
- Many are valuable pollinators
- Honey bees and bumblebees are eusocial or colonial
- Workers (sterile females) forage for nectar and pollen

Ex. Indian honey bee, *Apis cerana indica*
European honey bee, *Apis mellifera*
rock bee, *Apis dorsata*;
bumble bees, *Bombus* spp.;

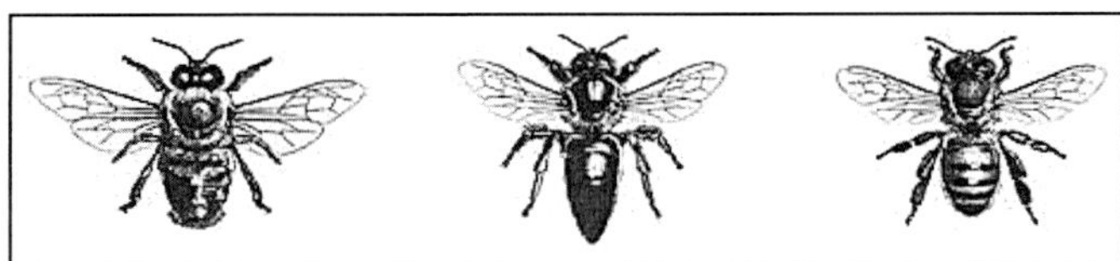

Family : Xylocopidae: (Carpenter bee)

- Pollen baskets are absent in hind legs but brushes are present.
- Build nests in dead logs and in live branches.
- They take nectar from flowers simply by biting through the base of the flower instead of sipping from the top.

carpenter bee, *Xylocopa* spp.

Family : Melliponidae

eg. Stingless bee, Melipona spp.

8

Collection and Preservation of Insects

Why Make an Insect Collection?

- An insect collection helps develop personal identification skills.
- An insect collection can be used to display insects or insect features so that others may learn about them.
- An insect collection can allow you to better observe details of insect structures and how they function.
- An insect collection can provide a record of when and where various insects occur.

Equipments required

- Forceps, vials containing alcohol or other preservatives
- Killing jars, Small boxes for storing specimens after their removal from killing jars.
- Aspirators
- Note book
- A strong knife, hand lens
- A small, fine brush for picking up minute specimens.

Methods used for collection of insects

1. **Hand picking:** It is a tedious practice for large insects like beetles and grasshoppers. It is not suitable for insects inflicting painful bites and stings.
2. **Insect net:** There are two types of insect nets.

i) Aerial net: (Butterfly net)

- The aerial net, sometimes called a "butterfly net," has a bag constructed of a light-weight mesh fabric that allows air to move readily through it and is fine enough to retain the insects.
- The net consists of three parts viz., hoop, handle and porous cloth bag made out of mosquito netting material. It has a small hoop (30-40 cm dia) and a long handle (100 cm).

- These nets are easily handled and can be used to capture fast flying insects.
- They also allow you to see the insects that you have captured, which is very useful when removing butterflies or stinging insects.
- However, aerial nets also are fairly delicate, and the mesh readily tears. Therefore, they are not good for sweeping vegetation.

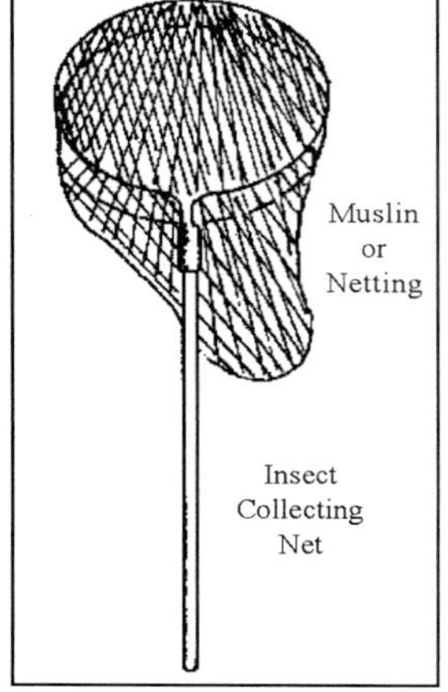

ii) Sweep net

- Sweep nets have a solid cloth sample bag made from muslin, feed-sack cloth, or some other fabric that resists ripping.
- Usually, the handle is stronger and heavier than that used in an aerial net, and the connection with the wire rim is more highly reinforced.
- These features allow the sweep net to sample grasses, crops, and shrubs, which can result in captures of large numbers of insects in a short time.

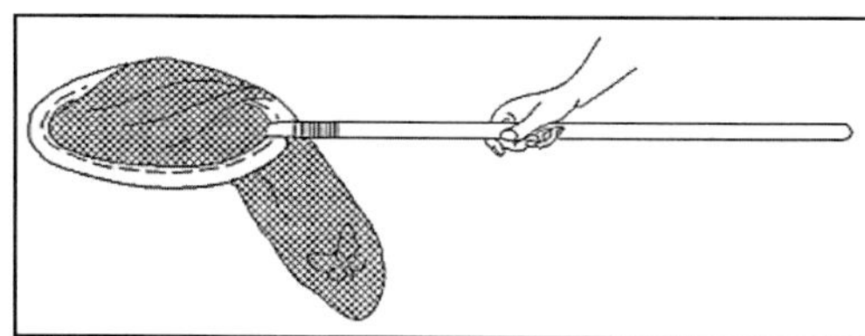

Fig: Twisting of insect net.

- This is suitable for collecting leaf hoppers, grasshoppers and other small insects. The net is swept over vegetation and quickly turned to fold the cloth bag over the hoop in order to prevent the escape of trapped insects.

iii) Aspirator (Pooster)

It is a device useful to collect all insects into a vial with no damage to the specimens. It is also useful for collecting insects from the insect net or any other surface.

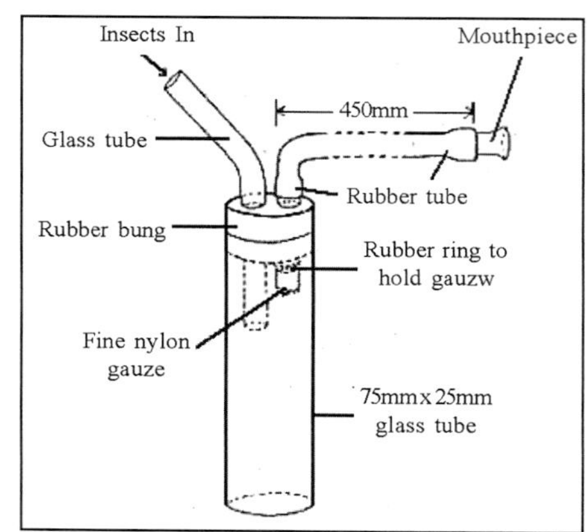

Fig: Aspirator

iv) Traps

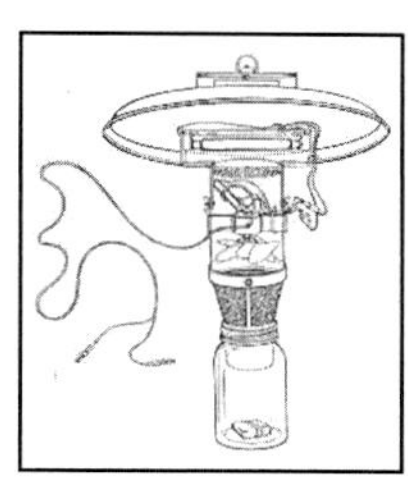
Light traps

Traps can be used for collecting different types of insects.

- Food lure trap-Flies
- Sex lure trap-Moths
- Water trap-Brown plant hopper
- Light trap-Positively phototropic insects
- Sticky trap-Whiteflies
- Suction trap-Airborne insect
- Malaise trap-Actively flying insect

Food lure trap

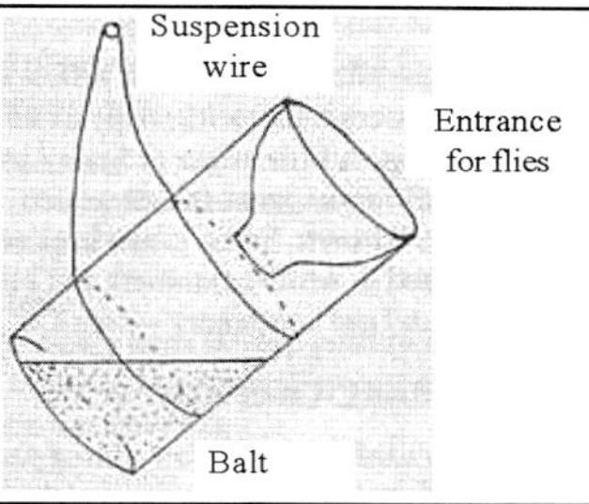

Pheromone trap

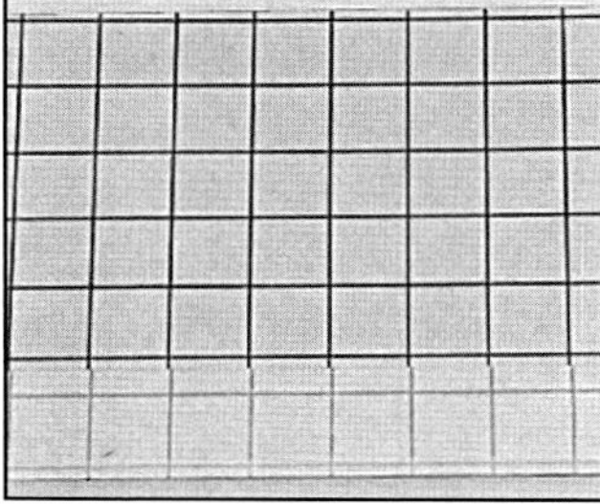
Sticky trap

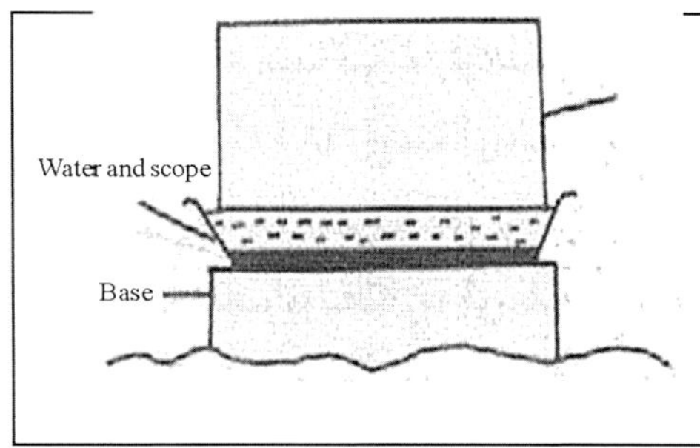

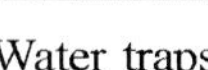
Water traps

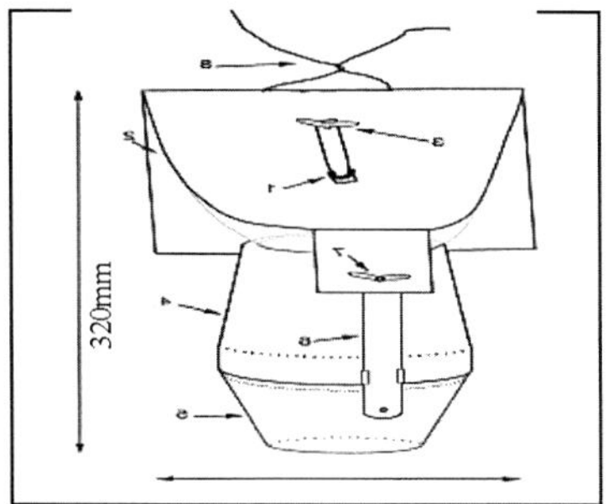

Funnel trap

v) Berlese funnel

Soil dwelling insects can be collected by using Berlese funnel.

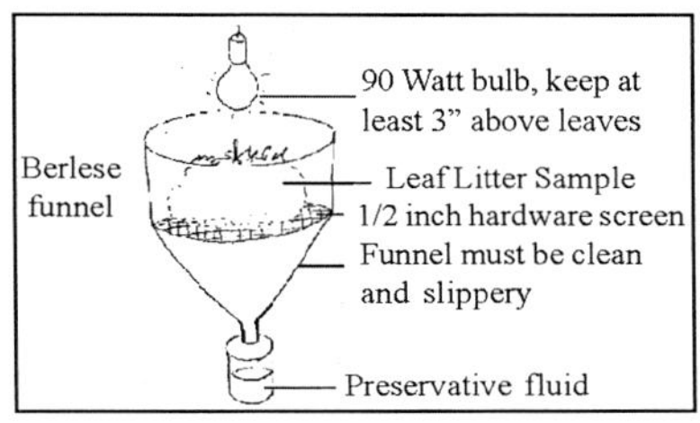

Berlese funnel

KILLING

- A killing jar is a very useful tool when making an insect collection. It allows one to rapidly kill any insects collected in the field so that they remain in good physical condition for use in your display.

- Ethyl acetate, found in nail polish remover, is probably the best killing agent to use in collections. When going on field trips, it sometimes is a good idea to bring along several killing jars.
- Specimens can be damaged if the killing jar gets overcrowded.
- Newly collected insects may scratch and tear the more delicate insects. Alternately, the wing scales of moths and butterflies killing should be done immediately after capture (thorax pinching).

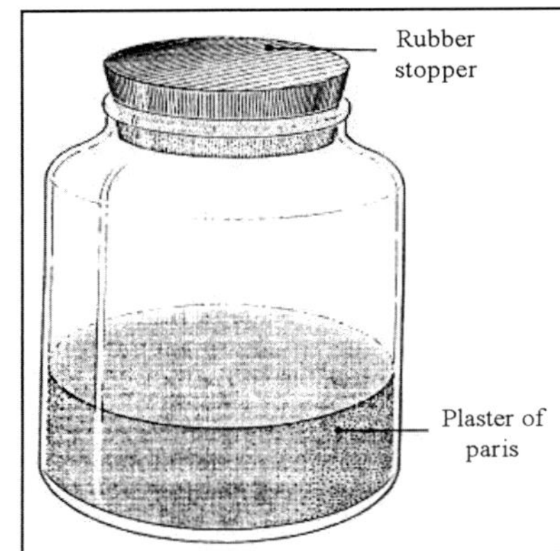

Killing jar with plaster of paris bottom

- Potassium cyanide, ethyl acetate, carbon tetra chloride (carbona) and chloroform are commonly used for killing insects. Potassium cyanide kills the insect quickly but rigor mortis sets in quickly. Ethyl acetate kills the insects more slowly and does not last long. But the dead insects remain in a relaxed condition for a longer time without becoming brittle.

Materials Needed

- An empty, wide-mouthed jar. (Peanut butter or Mayonnaise jars do well, but do not use plastic.)
- Plaster of Paris and Water
- A bottle of ethyl acetate (found in fingernail polish remover)
- A roll of adhesive tape and Sawdust or shavings.

Procedure for Making a Killing Jar

- Add about ½" saw dust or shavings to bottom of jar.
- Mix 8 heaping teaspoons of plaster of Paris with 5 teaspoons of water in a mixing cup.
- Stir the mixture until it is smooth then pour or spoon out the mixture into the bottom of the jar, over the saw dust.
- Gently tap the jar against the ground to make the plaster of Paris settle and make a smooth surface.
- Allow the Plaster of Paris to harden and dry for several days, with the cap off of the jar.

- Take the jar outdoors when adding the killing agent. Pour ethyl acetate (nail polish remover) onto the surface of the plaster. This will be absorbed into the bottom of the jar. Do not put so much in that the surface remains wet.

  Caution: Avoid inhaling much of the ethyl acetate.

Note

- You can put in a piece of tissue or some other cardboard to cover the plaster surface so that insects do not directly contact it.)
- Place the cap tightly on the jar. By keeping the cap on except when using it, enough of the killing agent will remain to keep it effective for several days. To recharge the jar, just add some more ethyl acetate to the plaster.
- Wrap the jar on the outside with adhesive tape to protect the killing jar from accidental breakage and reduce condensation in the jar, a particular problem when the jar is exposed to direct sunlight.
- Mark the jar prominently with the word Poison.

Pinching the thorax

A butterfly or moth can be immobilized and killed in an emergency by giving a sharp pinch on the thorax.

Killing with alcohol

Many insects can be killed by dropping them directly into 70 to 90% ethyl or isopropyl alcohol. (90% alcohol is used to store insect if the DNA of the insect to be extracted)

MATERIALS NEEDED FOR PRESERVATION

Paper folds (Paper envelopes)

They are useful for temporary preservation and storage of large winged insects such as dragonflies, butterflies or moths. These triangular envelopes can be made from a sheet of notebook or by using absorbent type of paper used in duplicating machines.

Cut the paper into rectangles with their sides in the proportion of 3:5. Bring the diagonally opposite comers together to leave two projecting flaps. Write the data regarding collection on the outer side of a projecting flap. Keep the immobilized insect in between the two overlapping triangles. Fold the flaps to produce a triangular envelope.

Setting board (Spreading board)

It is useful for spreading the wings of dead insects. It is a wooden board with a central groove in the middle. Flat cork strips are glued on either side of the groove and in the bottom of the groove to enable pinning. A thermocol sheet with a centrally cut groove can also be used as a substitute for the setting board.

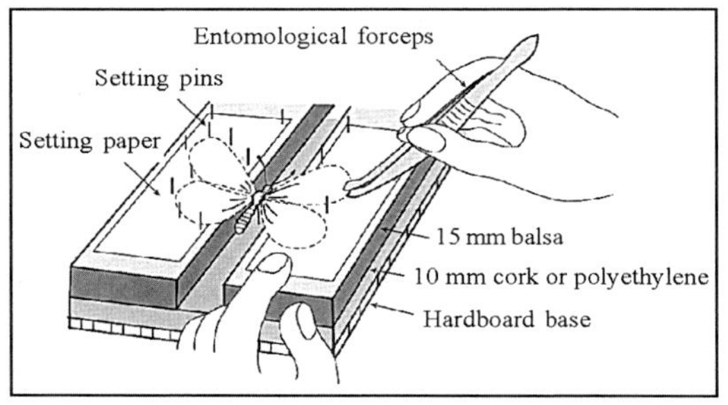

Setting board

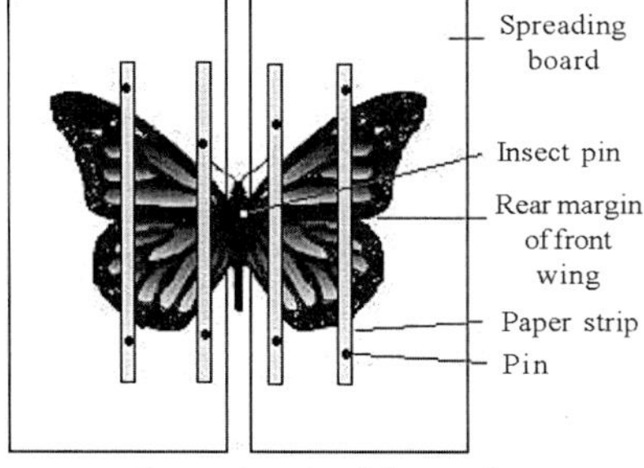

Properly pined butterfly

Relaxing container

Setting or mounting an insect should be done within a day after killing. Otherwise the insect will become stiff and brittle. Stiffness in the dead insect can be removed by placing it in a relaxing container. High humidity inside the relaxing container permits water to be reintroduced into the specimens thus making them flexible. Fill a container with sand to 1/4th of its capacity. Saturate the sand with water. Add a few drops of carbolic acid or formaldehyde to prevent mould growth. Keep the dried specimens in a small open box or in an uncovered petridish to avoid direct contact of the specimen with moist sand. Close the lid tightly and allow them to remain for a day or two until they become flexible.

Pins

Common pins are undesirable for pinning insects. Pins used for pinning insects should be slender, hard with a pointed tip and a small head. Pure nickel pins or nickel plated ones resist rusting. Commonly NO.16 and 20 pins are used for pinning larger and smaller insects respectively. In normal, non-stainless steel pins, a green substance deposited on the pin known as verdegris.

Micropins

For pinning very small insects micropins are used. They are very thin, slender, delicate and headless pins. They do not rust. They are also known as insect pins minuten pins or entomological pins.

Methods of preservation

1. Pinning

It is the best and most common method to preserve hard bodied insects. They will dry and remain in perfect condition on the pins without requiring any further treatment. During drying the outer exoskeleton remains intact while the inner soft tissues dry up. Insects can be pinned directly if they are big. They are pinned vertically through their body. During pinning the insect is held between the thumb and forefinger of one hand and the pin is inserted into the insect with the other hand.

While pinning 1/3rd length of the pin should be above the insect to permit a comfortable finger hold. Exact place of insertion of the pin varies among different groups of insects.

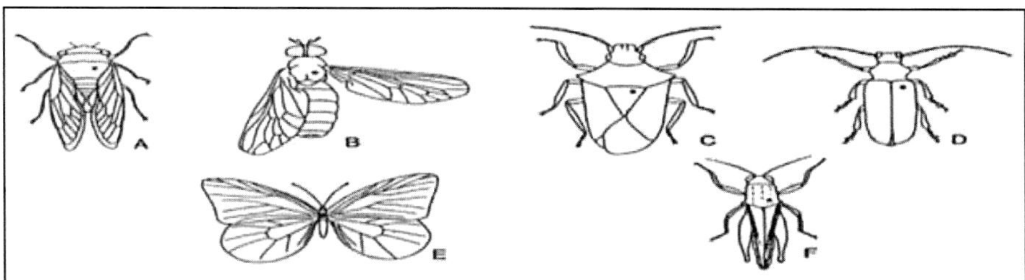

Proper location of pinning for : (A) cicada, (B) horsefly, (C) true bug, (D) beetle, (E) butterfly, and (F) grasshopper.

Common pinning position of insect

Pinning position	Insect /order
1. Pronotum	Grasshoppers, Crickets, Cockroaches, Preying Mantids,
2. Thorax	Dragonflies, Damselflies, Bees, Wasps. Butterflies, Moths and True Flies
3. Scutellum	Bugs
4. Right tegmina	Dermopteran
5. Metanotum	Stick and Leaf Insects
6. Right elytra	Beetles

2. Double mounting

Pinning is troublesome in smaller insects. Very small insects cannot be pinned because most of the body parts of the insects will be lost during pinning. For such insects double mounting can be followed.

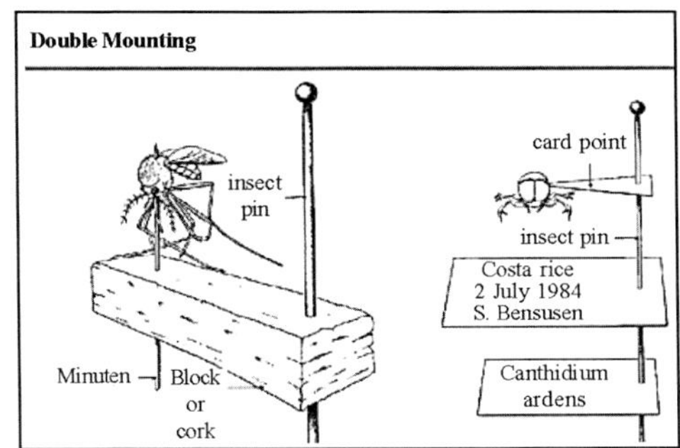

Staging

The stage is a narrow rectangular piece of pith or cork. The small insect is pinned correctly with a micro pin to the stage later the stage is pinned in the insect store box with a bigger pin.

Carding

A rectangular (5 x 8 mm or 5 x 12 mm) white card or celluloid bit may be used as stage. On the stage instead of pinning, the insect specimen is stuck on it by using transparent or stain free adhesive. A spot of good glue or white gum can also be used. The insect should not be embedded in the glue and only minimum quantity of the glue should be used. After mounting, the card is pinned to the insect storage box with a large pin.

Pointing

The insect specimen is glued to a card or celluloid bit into a triangle of 10 mm height and 5 mm base. Bend down the tip of the card to form a small surface to which the insect is stuck. Apply a drop of glue or adhesive by touching the point to the glue and to the thorax of the insect to be mounted. Press the right side of the specimen against angled and glued card tip. A bigger pin is inserted at the midpoint near the base for pinning the card with the insect to insect store box.

3. Liquid preservation

Soft-bodied forms (nymphs, larvae and many adults) shrivel when mounted dry. Such insects can be preserved in preservative fluids like ethyl alcohol (70%) and formalin (3-4 %). All these preservatives are highly volatile. Screw cap vials are satisfactory if the caps are tight fitting. Seal the stopper with paraffin wax and properly label.

4. Setting

Setting insects is essential to study the wing characters. It affords a better look to the preserved specimens. Wings of moths, butterflies, dragonflies and damselflies are set on either side. In grasshoppers, wings on one side alone are set. Setting boards are used for setting insects. Setting should be done before the insects become stiff.

LABELLING

Labels are must for every collection. Any collection should have a locality label giving particulars about date and locality of its capture. An additional label is often used that usually has the name or initials of the collector and the habitat or host from which the specimen is collected. Labels should be small, (12 x 6 mm) neat and made of stiff paper. Labels may be printed or hand written with micro tipped pen. They are inserted beneath the insects at 1/3rd height from the base. The long axis of the label should coincide with the long axis of the insect. If more than one label is used then the label should be parallel. All labels should be oriented so that they read from left side.

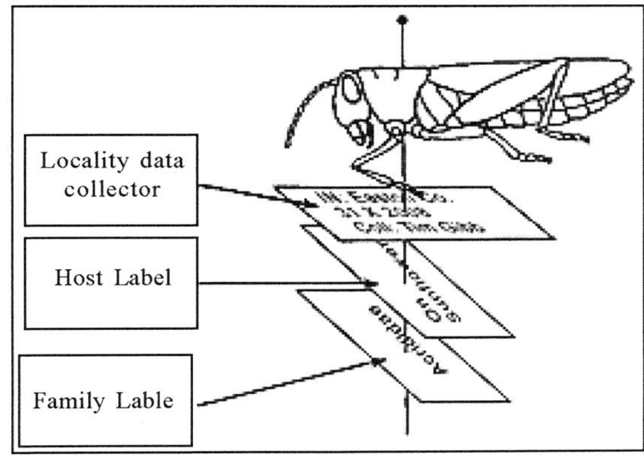

Procedures for lebelling

DISPLAY

Insect store boxes

Commonly wooden boxes of dimension 45 x 30 x 15 cm are used as insect store boxes for displaying preserved insects. The box should be light in weight, airtight and moisture proof with a well fitting hinged lid. A cell is provided inside to keep repellents. Cork sheets are glued to the inside of the top and bottom of the box to permit pinning. Glass topped boxes can be used for displaying insect collections but the colour of the preserved insects fades due to constant exposure to light.

Repellents and preservatives

Dermestid beetles, red flour beetle and psocids commonly attack preserved specimens. Naphthalene balls mounted on pins are pinned inside to repel museum insects. This is done by heating the head of a pin in flame and pressing it against a naphthalene ball. Naphthalene flakes can also be kept in perforated envelopes and can be pinned in the boxes. In the place of naphthalene, para-dichloro-benzene (PDB) crystals can be used.

9

Objective Questions

9.1 Multiple Choice Questions

1 Which insect order is close related to Diptera?

a) Hymenoptera b) Orthoptera
c) Plecoptera d) Zygentoma

2 Lobsters and shrimp belong to which class?

a) Crustacea b) Arachnida
c) Myriapoda d) Xiphosura

3 Which insect comes in order Hemiptera?

a) Bed bugs and stink bugs b) Chewing and sucking lice
c) Roaches and mantids d) Crickets and grasshoppers

4 Which order is do comes under holometabolous?

a) Siphonaptera b) Hymenoptera
c) Thysanoptera d) Neuroptera

5 Which order is mainly herbivorous?

a) Trichoptera b) Odonata
c) Phasmatodea d) Thysanoptera

6 Which order is mostly parasitic?

a) Diplura b) Siphonaptera
c) Zoraptera d) Diptera

7 Sucking mouthparts are not found in:

a) Fleas b) Lice
c) Flies d) Ants

8 All ametabolous insects are:

a) Predatory b) Wingless
c) Endognathous d) All of these

9 Immatures of the Neuroptera would be classified are:

a) Scavengers b) Parasites

c) Herbivores d) Predators

10 Which characters of orders Mantodea, Dermaptera, and Isoptera have common?

a) Winglessness b) Chewing mouthparts

c) Herbivory d) All of these

11 Which of the following insect order is more close to Dermaptera?

a) Hymenoptera b) Orthoptera

c) Plecoptera d) Zygentoma

12 Which arthropods have chelicerate type mouth part?

a) Spiders b) Millipedes

c) Shrimp d) All of these

13 The order Orthoptera contains ____________.

a) Bed bugs and stink bugs b) Chewing and sucking lice

c) Roaches and mantids d) Crickets and grasshoppers

14 Which characteristic feature is not found in the Onychophora?

a) One pair of antennae b) Three tagmata

c) Jointed legs with claws d) Segmented body

15 Which order are commonly blood feeders?

a) Siphonaptera b) Thysanoptera

c) Phasmida d) Hymenoptera

16 Which structure is related to the Hymenoptera?

a) Furcula b) Hamuli

c) Collophore d) Elytra

17 Chewing mouthparts is not present in:

a) Fleas b) Earwigs

c) Beetles d) Bees

18 All neopterous insects are:

a) Predatory b) Wingless

c) Ectognathous d) Hemimetabolous

19 Which developmental stage is related with Ephemeroptera?

a) Subimago c) Naiad

c) Imago d) All

20 Which orders are most important in the transmission of human disease?

a) Phasmatodea and Odonata

b) Hymenoptera and Psocodea

c) Diptera and Siphonoptera

d) Hemiptera and Thysanoptera

21 a) naiad is best described as:

a) Predatory larva b) Wingless adult

c) Aquatic nymph d) Scavenger

22 Chelicerate arthropods include:

a) Millipedes and centipedes b) Lobsters and shrimp

c) Spiders and ticks d) Lice and fleas

23 Which statement is related to all crustacea?

a) They live on land

b) They have chewing mouthparts

c) They have six walking legs

d) They are paleopterous

24 Which insect order is not associated with plants?

a) Hymenoptera b) Thysanoptera

c) Hemiptera d) Siphonaptera

25 The pathogen cause Human diseases are transmitted by which order?

a) Hymenoptera b) Thysanoptera

c) Diptera d) All of these

26 Odonata and Plecoptera are more similar to each other because both have:

a) Aquatic nymphs b) Endopterygote development

c) Paleopterous wings d) All of these

27 Which insect order not present in aquatic habitate?

a) Trichoptera b) Plecoptera

c) Diptera d) Orthoptera

28 Hemiptera and Hymenoptera are similar because both have:

a) Holometabolous development

b) Piercing-sucking mouthparts

c) Neopterous wings

d) All of these

29 Which insect order is most commonly found in soil litter?

a) Collembola b) Neuroptera

c) Lepidoptera d) Phasmatodea

30 In Isoptera winged forms are called:

a) Primary reproductives b) secondary reproductives

c) Tertiary reproductives d) none of the above

31 Silk glands are present in ______________.

a) Dermaptera b) Isoptera

c) Plecoptera d) Psocoptera

32 Hemimetabolous type of development is seen in

a) Psocoptera b) Plecoptera

c) Dermaptera d) Isoptera

33 Swarming is seen in

a) Dermaptera b) Plecoptera

c) Isoptera d) Pthiraptera

34 In Dermaptera parental care is shown by

a) Male earwigs b) Female earwigs

c) Both a & b d) None

35 Which of the following are only hemimetabolous insects that exhibit social behavior?

a) Termites b) Stone flies

c) Earwigs d) Booklice

36 In which of the fallowing polymorphism is seen:

a) Plecoptera b) Isoptera

c) Psocoptera d) Dermaptera

37 In which caste of Isoptera frontal gland is present

a) Workers b) Soldiers

c) Queen d) King

38 Cannibalism is seen in

a) Isopteran b) Phthiraptera

c) Plecoptera d) Dermaptera

39 Moniliform antennae found in

a) Plecoptera b) Isoptera

c) Dermaptera d) Psocoptera

40 Physapoda is another name of

a) Isoptera b) Coleoptera

c) Thysanoptera d) Diptera

41 Spurious veins are present in

a) Robber flies b) Hover flies

c) Both a & b d) None

42 Eye flies "*Siphunculina funicola*" belongs to the family

a) Psychodidae b) Chloropidae

c) Drosophilidae d) Muscidae

43 Which of the following order closely related to Lepidoptera is

a) Strepsiptera b) Diptera

c) Trichoptera d) Neuroptera

44 Tail like prolongation of hind wings in papalionidae is

a) M_1 vein b) M_2 vein

c) M_3 vein d) All the above

45 Mandibular articulation in head is monocondylic

a) Apterygota b) Pterygota

c) Both d) None

46 The insect group have larval pupal stages in their life cycle

a) Paleopteran order 0 b) Panorpoid complex

c) Orthopteroid orders d) None

47 Housefly belongs to sub orders

a) Nematocera b) Brachycera

c) Cyclorrapha d) All

48 The chemical present in fireflies

a) Luciferin b) Isoluciferin

c) Both d) None

49 The tarsal formula for chrysomelidae is

a) 3-3-3 b) 4-4-4

c) 5-5-5 d) 5-5-4

50 According to the 'ICZN, the ending "Oidea" is for

a) Superfamily b) Subclass

c) Subfamily d) Suborder

51 The order which establishes the relationship of insects with other arthropods is

a) Plecoptera b) Thysanura

c) Collembola d) Diplura

52 Which of the following are bark beetles

a) Scymnus b) Scolytus

c) Daucus d) Myllocerus

53 The embioptera the silk glands and spinnerets found in

a) Basitarsus of forelegs b) Basitarsus of middle legs

c) Pretarsus of forelegs d) Pretarsus of middle legs

54 Which organization published Zoological Record

a) BIOSIS

b) European Society of Entomology

c) British Society of India

d) None

55 Tarsus single segmented in case of

a) Clasping type b) Fossorial type

c) Scansorial type d) Natatorial type

56 Which Arthropods have chelicerae

a) Millipedes b) Spiders

c) Shrimp d) Onychophora

57 Which order has both holo & Hemimetabolous insects

a) Hymenoptera b) Diptera

c) Strepsiptera d) Siphenaptera

58 Structurally and functionally evolved order is

a) Mecoptera b) Hymenoptera

c) Diptera d) Coleoptera

59 How many pairs of legs are present in the class Aracnida

a) 5 pairs b) Two pairs

c) Four pairs d) Three pairs

60 Filter chamber is present in ______.

a) Heteroptera b) Homoptera

c) Hymenoptera d) Thysanoptera

61 Which family has 5 segmented antenna? ______.

a) Acrididae b) Nepidae

c) Pentatomidae d) Phasmidae

62 Repugnatorial glands open in which part of the leg_____.

a) Tibia b) Trochanter

c) Femur d) Coxa

63 Which of the following family has Corium, clavus, cuneus in its wing but lacks embolium?

a) Lygaeidae b) Anthocoreidae

c) Miridae d) Delphacidae

64 Aphid sucks the food from _____.

a) Xylem b) Phloem

c) Epidermis d) None of these

65 Hind tibia with two row of spines is found in ______.

a) Cicadellidae b) Delphacidae

c) Membracidae d) Lophopidae

66 Hind tibia with one large mobile flattened spur is found in _____.

a) Membracidae b) Delphacidae

c) Coccidae d) Cicadellidae

67 Hind wings are reduced to halteres in ______.

a) Butterfly b) Honeybee

c) Housefly d) Grasshopper

68 Whitefly classification is based on the structure of ______.

a) Larva b) Pupa

c) Egg d) None of these

69 Link between Hemipteroid & Endopterygota is ________

a) Thysanura b) Thysanoptera

c) Zoraptera d) Embioptera

70 Aquatic naiads are present in ____________
a) Ephemeroptera b) Odonata
c) Orthoptera d) Both a & b

71 Wing pads of nymphs undergo reversal during development of
a) Ephemeroptera b) Odonata
c) Orthoptera d) Both b & c

72 Scolopophores are the part of
a) Mullars organ b) Palmers organ
c) Comstock kellogs gland d) None of the above

73 Dicoptic eyes present in
a) Anisoptera b) Zygoptera
c) Both d) None of the above

74 Adult Odonata lacinia & galea fused and form
a) Mala b) Mask
c) Gula d) Papillae

75 Family comes under suborder caelifera
a) Acrididae b) Gryllidae
c) Gryllotalpidae d) Tettigonidae

76 In phasmidae line of weakness is present in _______ part of leg segment
a) Femur b) Tibia
c) Trochanter d) Tarsus

77 Synonyms of Ephemeroptera
a) Plecoptera b) Ephemerida
c) Myrientomata d) Perlaria

78 Mantidae are
a) Phatophagus b) Predators
c) Parasitoids d) All the above

79 Wing flexing mechanism absent in ________
a) Odonata b) Ephemeroptera
c) Both a & b d) None

80 Both wings and antennae are absent in
a) Ectognathous b) Protura
c) Collembola d) Dipluran

81 Collophore is present in the ________ abdominal segment of Collembola)

a) 3rd b) 1st

c) 4th d) 2nd

82 Anamorphic development is seen in

a) Thysanura b) Protura

c) Collembola d) Dipluran

83 Size of collembolans is

a) <2mm b) < 4mm

c) < 6mm d) < 8mm

84 Abdomen is 6 segmented in

a) Collembola b) Proturan

c) Dipluran d) Thysanura

85 Size of proturans

a) <4mm b) <6mm

c) <2mm d) < 8mm

86 Piercing type of mouth parts seen in

a) Collembola b) Dipluran

c) Proturan d) Thysanura

87 Limb like appendages called styli are present in 1st three abdominal segments of

a) Protura b) Collembola

c) Diplura d) Thysanura

88 In collembola, collophore,tenaculum and furcula are present on________ abdominal segments.

a) 2, 3, 4 b) 1, 3, 4

c) 2, 4, 5 d) 1, 3, 6

89 Endognathous mouthparts are present in

a) Diplura b) Collembolan

c) Protura d) All

90 The only hymenopteran pest attacking crop is –

a) Bethylid wasp b) Leaf cutter bee

c) Carpenter bee d) Mustard saw fly

91 Mouthpart of Hymenopteran pest except mustard saw fly is-

a) Cutting and chewing
b) Chewing and lapping
c) Mandibusuctorial
d) Siphoning type.

92 Prolegs lacks crochets in the larvae of-

a) *Spodoptera litura*
b) *Helicoverpa armigera*
c) *Athalialeugens proxima*
d) Both a & b

93 Mandibles are scissors like in-

a) Megachilidae
b) Xylocopidae
c) Apidae
d) Vespidae

94 Male apterous in –

a) Evaniidae
b) Aganidae
c) Sphecidae
d) Vespidae

95 Wing type of Honey bee is-

a) Tegmina
b) Elytra
c) Membranous
d) Hemi elytra

96 Highly evolved insect order is-

a) Dipreran
b) Hymenoptera
c) Lepidoptera
d) Coleopteran

97 Ovipositor longer than the body of itself is the character of –

a) Ichenumonoid wasp
b) Braconidae
c) vespidae
d) Bethylid wasp

98 The word "Hymen" in Hymenoptera means

a) Thin
b) Thick
c) Vein
d) Membrane

99 Erusiform larvae present in-

a) Apidae
b) Trenthitinidae
c) Platygasteridae
d) Formicidae

100 Prothorax is enlarged and shield-like pronotumin ---------------

a) Orthoptera
b) Blattodea
c) Mantodea
d) Phasmatodea

101 Cerci is short and many segmented in ---------------

a) Mayfly
b) Cricket
c) Cockroach
d) Grasshopper

102 Jumping or leap cockroach belongs to the family ——————

a) Acrididae b) Tettogonidae

c) Corydidae d) Cryptoceracidae

103 Head of Grylloblattodea insects is of —————— type

a) Prognathous b) Hypognathous

c) Opisthognathous d) None of these

104 Mouthparts in mantaphasmatids are ——————

a) Chewing and lapping type b) Rasping and sucking type

c) Piercing and sucking type d) Chewing and biting type

105 In embiids silk glands are located on the ——————

a) Fore tibia b) Basitarsi of foreleg

c) Hind femur d) Basitarsi of hindleg

106 Cannabalism is seen in ——————

a) Orthoptera b) Zoraptera

c) Grylloblattodea d) Embioptera

107 Zorapterans are commonly called as ——————

a) Snowfleas b) Gladiators

c) Angel Insects d) Web Spinners

108 Males have asymmetrical genitalia in case of ——————

a) Mantodea b) Grylloblattodea

c) Blattodea d) Mantaphasmatodea

109 —————— resemble both crickets and cockroaches

a) Mantaphasmatodea b) Grylloblattodea

c) Zoraptera d) Phasmida

110 Which family have complete metamorphosis

a) Neuroptera b) Megaloptera

c) Trichoptera d) All the above

111 Neuropteran flies are called —————— fliers

a) Weak b) Strong

c) Lazy d) None of these

112 Larvae of neuropterans, having —————— type of mouth parts

a) Chewing b) Sucking

c) Mandibulosuctorial d) None of these

113 Stalked eggs are produced by ————————

a) Antlion b) Greenlace wing fly

c) Owlflies d) None of these

114 In neuroptera, which family having hypermetamorphosis ——————

a) Ascalaphidae b) Myrmeleontidae

c) Mantispidae d) Chrysopidae

115 Snake flies coming under which order ————————

a) Mecoptera b) Megaloptera

c) Trichoptera d) Rhaphidioptera

116 Conical pits are constructed by ————————

a) Antlions b) Mantispidfly

c) Scorpionfly d) Caddisfly

117 Siphonaptera is derived from which word ————————

a) Latin b) Greek

c) Latin & Greek d) None of these

118 Roof like wing in the orders ————————

a) Neuroptera b) Trichoptera

c) Both a & b d) None of these

119 Wing tracheation is totally absent in this order ————————

a) Mecoptera b) Trichoptera

c) Both a & b d) None of these

120 Cerci and distinct ovipositor absent in which order?

a) Coleopteran b) Blattodea

c) Lepidoptera d) Odonata

121 In Gyrinidae middle leg and hind leg modified into

a) Saltatorial b) Fursorial

c) Cursorial d) Natatorial

122 Usually coccinellids are which type of larva?

a) Compodeiform b) Eruciform

c) Apodus d) Both a & b

123 Photogenic organ present in which abdominal segement s in case of lampiridae?

a) 1 & 2 b) 4 & 5

c) 6 & 7 d) 7 & 8

124 Luminescence in fire fly due presence of which enzyme?

a) Luciferin b) Oxy luciferin

c) Luciferacin d) Luciferase

125 Family of furniture beetle/

a) Bostrychidae b) Cerambicidae

c) Cassidae d) Meloidae

126 Fore wing of coleopteran called as?

a) Modified wing b) Elytra

c) Hemelytra d) Sclirites

127 Damaging stage of sweet potato weevil is?

a) Grubs b) Adult

c) Both a and b d) None of the above

128 Family of meal worm is?

a) Melolontidae b) Tenebrionidae

c) Galerucidae d) Meghachilidae

129 The word Coleo means?

a) Membrane b) Sheath

c) Wing d) Sclerites

130 Labella is present in-

a) Culicidae b) Muscidae

c) Tachinidae d) Tephritidae

131 Anthrax is spread by-

a) Asilidae b) Tabanidae

c) Culicidae d) Muscidae.

132 spurious vein present in

a) Tabanidae b) Asilidae

c) Syrpidae d) Drosophilidae

133 chest bone present in

a) Prothorax b) Mesothorax

c) Metathorax d) Abdomen.

134 Diptera subdivided into ———————— suborder.

a) 2 b) 3

c) 4 d) 5

135 The order having totally parasite insect.

a) Dermoptera b) Zoroptera

c) Embioptera d) Pthiraptera

136 Which of the following insect adult is over wintering stage.

a) May fly b) Ear wig

c) Stone fly d) Damsel fly

137 In cricket the tympanum organ situated in

a) 1st abdominal tergum b) 2nd abdominal tergum

c) Fore tibiae d) Hind tibiae

138 Cerci absent in order.

a) Psocoptera b) Phasmid

c) Zoroptera d) Dermaptera

139 Unsegmented cerci and leathery fore wing is characteristic feature.

a) Orthoptera b) Hemiptera

c) Psocoptera d) Diplura

140 Fore leg raptorial type

a) Reduviidae b) Mantidae

c) Napidae d) Both b) and c)

141 Scent gland of coreidae opening on

a) Metathorax b) Prothrox

c) Mrsothorax d) Abdomen

142 Scent gland of cimicidae present on

a) Metathorax b) Prothrox

c) Mrsothorax d) 1-3 abdomen segment

143 The correct sequence of taxonomic study are

a) Classification, Identification, Characterization, Nomenclature

b) Characterization, Classification, Identification, Nomenclature

c) Characterization, Identification, Nomenclature, Classification

d) None

144 The term systematic given by

a) Linnaeus b) Aristotle

c) Whittake d) None

145 Systematic is the term used to refer:

a) Phenetics + Taxonomy b) Phylogenetic + Biology

c) Phylogeny + Taxonomy d) Dendrogram + Biology

146 The document that includes all the information related to a particular genus or family is termed as:

a) Monograph b) Record

c) Revision d) Animal module

147 Which one of the following is NOT covered under Taxonomy?

a) Alpha taxonomy b) Beta taxonomy

c) Deltataxonomy c) Gamma taxonomy

148 Name the philosopher who first classified organisms?

a) Whittekar b) Carl woese

c) Linnaeus d) Aristotle

149 Name the organization which provides rules for naming animals.

a) ICZN b) ICN

c) ICBN d) IBM

150 Who is considered as the father of taxonomy?

a) Aristotle b) Linnaeus

c) Whittaker d) Earnst Haeckel

151 What is tautonym?

a) These are the repeated sequences b) It is a name of fish

c) Identical name of genus and species c) It is a name of the genus

152 Arrange in the correct order. 1) Class, 2) Kingdom, 3) Phylum, 4) Order, 5) Genus, 6) Family, 7) Species

a) 6, 2, 4, 1, 5, 7, 3 b) 7, 1, 3, 4, 5, 6, 1

c) 1, 2, 3, 4, 5, 6, 7 d) 2, 3, 1, 4, 6, 5, 7

153 What is the term given to a duplicate specimen of original type?

a) Lectotype b) Holotype

c) Isotype d) Neotype

154 Interaspecific category of similar genetic composition for biological attributes is called

a) Race b) Deme

c) Biotype d) None

155 Intermediate population unit which exist between species and deme is called

a) Race b) Family

c) Biotype d) Order

156 Tribe present between

a) Sub family and genus b) Order and family

c) Kindom and phyllum c) None

157 DNA barcoding in insects is carried out by using :

a) Nuclear gene b) Mitochondrial gene

c) tRNA d) mRNA

Answer key

1	**2**	**3**	**4**	**5**	**6**	**7**	**8**	**9**	**10**
a	a	a	c	c	c	d	b	d	b
11	**12**	**13**	**14**	**15**	**16**	**17**	**18**	**19**	**20**
c	a	d	b	a	b	a	c	d	c
21	**22**	**23**	**24**	**25**	**26**	**27**	**28**	**29**	**30**
c	c	b	d	c	a	d	c	a	a
31	**32**	**33**	**34**	**35**	**36**	**37**	**38**	**39**	**40**
d	b	c	b	a	b	b	d	b	c
41	**42**	**43**	**44**	**45**	**46**	**47**	**48**	**49**	**50**
b	b	c	c	a	b	c	a	c	a
51	**52**	**53**	**54**	**55**	**56**	**57**	**58**	**59**	**60**
d	b	a	a	c	b	c	b	c	b
61	**62**	**63**	**64**	**65**	**66**	**67**	**68**	**69**	**70**
c	d	c	a	a	b	c	b	b	d
71	**72**	**73**	**74**	**75**	**76**	**77**	**78**	**79**	**80**
d	a	b	a	a	c	b	b	c	b
81	**82**	**83**	**84**	**85**	**86**	**87**	**88**	**89**	**90**
b	b	c	a	c	c	a	b	d	d
91	**92**	**93**	**94**	**95**	**96**	**97**	**98**	**99**	**100**
b	c	a	b	c	b	a	d	b	b
101	**102**	**103**	**104**	**105**	**106**	**107**	**108**	**109**	**110**
c	d	a	d	b	d	c	b	b	d
111	**112**	**113**	**114**	**115**	**116**	**117**	**118**	**119**	**120**
a	c	b	c	d	a	b	c	c	a
121	**122**	**123**	**124**	**125**	**126**	**127**	**128**	**129**	**130**
d	a	c	d	a	b	c	b	b	b
131	**132**	**133**	**134**	**135**	**136**	**137**	**138**	**139**	**140**
b	c	a	b	d	b	c	a	a	d
141	**142**	**143**	**144**	**145**	**146**	**147**	**148**	**149**	**150**
a	d	b	a	c	a	c	d	a	b
151	**152**	**153**	**154**	**155**	**156**	**157**			
c	d	d	c	a	a	b			

9.2 True/False

1. Spiders, ticks, and horseshoe crabs belong to the subphylum Mandibulata.
2. Dragonflies are more closely related to scorpionflies than to stoneflies.
3. The immature stages of some holometabolous insects are called naiads.
4. Insects in the orders Orthoptera and Hymenoptera have chewing mouthparts.
5. Both Arthropods and Onychophorans have open circulatory systems.
6. Centipedes and millipedes belong to the SAME subphylum as spiders and ticks.
7. Honey bees are more closely related to ants than to butterflies and moths.
8. Aquatic immatures of ALL holometabolous insects are known as larvae.
9. All insects in the orders Diptera and Psocodea have sucking mouthparts.
10. Cerci are commonly found in ametabolous insects but NOT in holometabolous insects.
11. All arthropods have jointed legs with claws.
12. All ametabolous insects are apterygote.
13. All orthopteroid orders are hemimetabolous.
14. All holometabolous orders are endopterygote.
15. All chelicerate arthropods are aquatic.
16. In Mastotermitidae family of Isoptera true workers are absent.
17. Stone flies are serving as ecological indicators.
18. Exchange of food between termites is known as Trophyllaxis.
19. Cerci are present in psocoptera.
20. Termites are having rasping and sucking type of mouthparts.
21. In Kalotermitidae true workers are present.
22. In Isoptera, termite taxa without soldiers is Anoplotermis and Apeculitermis
23. Pterostigma present on the wings of book lice.
24. Distinct Y-shaped suture is absent in the order Psocoptera
25. Hind wings are semicircular, ear like in Dermaptera
26. Gnats is example of diptera.
27. All thoracic segments are fused together in diptera.
28. Pupa of brachycera is coarctate.

29. In culicidae anal gill presents in terminal segment of abdomen.
30. In tabanidae male eye is dichoptic.
31. Adult of syrphidae helps in pollination.
32. Life cycle of drosophilidae is 21-28 days.
33. Hippoboscidae is viviparous.
34. First abdomenal segment of musicidae is brown.
35. Haltere is organ of equilibrium.
36. Siphonaptera having roof like wings.
37. Trichopteran insects look like moths.
38. Green lacewing act as a predator.
39. Mantispidflies are same like as preying mantids.
40. Antlions are same like as dragonfly.
41. Scorpion fly coming under Megalopteran order.
42. Wing tracheation is present in trichoptera.
43. Siphonaptera having complete metamorphosis.
44. Snakeflies having pterostigma in wing region.
45. Green lacewing having golden yellow color of eye.
46. Eggs are laid on the back of the male in belastomatid.
47. Cornicles are present on 4th and 5th abdominal segment.
48. Ovipositor is saw like in tubulifera.
49. Left mandible is absent in thysanoptera.
50. Forewings are hemielytra in homoptera.
51. Gular region is well defined in Heteroptera.
52. Cicadas belong to the family cicadellidae.
53. Caudal breathing tube is absent in Nepidae.
54. Giant water bug is herbivores in nature.
55. Plant hoppers belongs to the family Cicadellidae.
56. Most of scarabids are defoliater in nature.
57. The antenna type of female *callosobruchus chinensis* is unipectinate.
58. The exposed parts of abdomen in bruchidae family referred as **Pygidium**.
59. Pupation of pulse beetle occurs outside the seed.

60. *Cylas formicarius* attack only in storage.
61. Flat headed borers are good pollinators in groundnut crop.
62. Cerambycidae family insects are usually called **Round headed borer**.
63. Antenna type of *Myllocerous* species is geniculate?
64. Tortoise beetle is belongs to family cassididae?
65. Horn is longer and curved in female dynastid beetle.
66. Cockroaches are holometabolous in development
67. Ice crawlers are primarily nocturnal.
68. Mantaphasmatids are found in moderately humid to very dry habitats.
69. Grylloblattodeans are closely related to cockroaches and stick insects.
70. Ocelli are absent in embiids.
71. In embiids, only females spin silk.
72. Rapid rearward movement of webspinners is enabled through large depressor muscles in the hind tibia.
73. Only adults of embiids live in silken galleries.
74. Wing coupling mechanism of hymenoptera is Jugate type.
75. Mustard saw flies have 5 pairs of prolegs.
76. Eulophidae are mostly pupal parasitoids.
77. Abdomen is petiolated in Apocrita.
78. Fig wasp comes under the family Agaonidae.
79. Waggle dance and round dance present in hornet wasp.
80. Family of *Evania apendigaster* is Agonidae.
81. Nuptial flight is common for both Termites and honey bee.
82. Coccon made with mud is common feature of Apidae.
83. Saw flies have hatlere in the place of Hindwings.
84. Malphigian tubules are represented as papillae in proturan.
85. All Apterygotes are primitively wingless.
86. Forelegs function as antennae in Collembola.
87. Antenna and tentorium absent in Protura.
88. Thysanura posses entognathous mouthparts.
89. In thysanura compound eyes meet at the top of head.
90. Ventral tube of collembolan located on 3rd abdominal segment.

91. Proturans are flightless insects with entognathous mouthparts.
92. Collophore in collembolans act as jumping organ.
93. Collembola is mostly found in soil litter.
94. Silken gland of Embioptera present on hind tibae.
95. Right mandible absent in order thysanura.
96. Prehensile organ is modification of mandible in odonata.
97. Anal lobe absent in psocoptera.
98. Jumping bristle belong to order Archeognatha.

Answer key

1	**2**	**3**	**4**	**5**	**6**	**7**	**8**	**9**	**10**
F	F	F	T	T	F	T	F	F	T
11	**12**	**13**	**14**	**15**	**16**	**17**	**18**	**19**	**20**
T	T	T	T	F	T	T	T	F	F
21	**22**	**23**	**24**	**25**	**26**	**27**	**28**	**29**	**30**
F	T	T	F	T	T	T	F	T	F
31	**32**	**33**	**34**	**35**	**36**	**37**	**38**	**39**	**40**
T	F	T	T	T	F	T	T	F	F
41	**42**	**43**	**44**	**45**	**46**	**47**	**48**	**49**	**50**
F	F	T	T	T	T	F	F	F	F
51	**52**	**53**	**54**	**55**	**56**	**57**	**58**	**59**	**60**
T	F	F	F	F	F	F	T	F	F
61	**62**	**63**	**64**	**65**	**66**	**67**	**68**	**69**	**70**
F	T	T	T	F	F	T	T	T	T
71	**72**	**73**	**74**	**75**	**76**	**77**	**78**	**79**	**80**
F	F	T	F	F	T	T	T	F	F
81	**82**	**83**	**84**	**85**	**86**	**87**	**88**	**89**	**90**
T	F	F	T	T	F	T	F	T	F
91	**92**	**93**	**94**	**95**	**96**	**97**	**98**	**99**	**100**
T	F	T	F	F	F	T	T		

9.3 MATCHING

1. Match

1	Cricket	a.	Hemiptera
2	Damselfly	b.	Hymenoptera
3	Antlion	c.	Odonata
4	Sawfly	d.	Neuroptera
5	Aphid	e.	Orthoptera

Ans. 1.e, 2.c, 3.d, 4.b, 5.a

2. Match

1	Odonata	a.	Jumping Bristles
2	Coleoptera	b.	Stonefly
3	Mecoptera	c.	Weevil
4	Plecoptera	d.	Scorpionfly
5	Archeognatha	e.	Dragonfly

Ans. 1. e, 2.c, 3.d, 4.b, 5.a

3. Match

1	Silverfish	a.	Plecoptera
2	Stoneflies	b.	Mecoptera
3	Scorpionflies	c.	Trichoptera
4	Caddisflies	d.	Orthoptera
5	Katydids	e.	Thysanura

Ans. 1. e, 2.a, 3.b, 4c, 5.d

4. Match

1	Elytra	a.	Collembola
2	Subimago	b.	Coleoptera
3	Furcula	c.	Ephemeroptera
4	Ootheca	d.	Collembola
5	Hamula	e.	Blattodea

Ans. 1. b, 2.c, 3.a, 4.e, 5.a

5. Match

1	Springtails	a.	Hemiptera
2	Leafhoppers	b.	Archeognatha
3	Bristletails	c.	Psocodea
4	Booklice	d.	Orthoptera
5	Crickets	e.	Collembola

Ans. 1. e, 2.a, 3.b, 4c, 5.d

6. Match

1	Siphonaptera	a.	Predators
2	Odonata	b.	Parasites
3	Thysanoptera	c.	Herbivores
4	Phasmatodea	d.	Scavengers
5	Mecoptera	e.	Phytophage

Ans. 1. b, 2.a, 3.e, 4.c, 5.d

7. Match

1	Pearmans organ	a.	Hodotermitidae
2	Perlidae	b.	Dermaptera
3	Forficulidae	c.	Psocoptera
4	Termitidae	d.	Plecoptera
5	Rottenwood termites	e.	Isoptera

Ans. 1. c, 2.d, 3.b, 4.e, 5.a

8. Match

1	Earwigs	a.	Phthiraptera
2	Termites	b.	Plecoptera
3	Book lice	c.	Dermaptera
4	Parasitic lice	d.	Psocoptera
5	Stone flies	e.	Isoptera

Ans. 1. c, 2.e, 3.d, 4.a, 5.b

9. Match

1	Cicadellidae	a.	Leaf footed bug
2	Coreidae	b.	Bedbug
3	Membracicdae	c.	Leafhopper
4	Cimicidae	d.	Tree hopper
5	Nepidae	e.	Water scorpion

Ans. 1. c, 2.a, 3.d, 4.b, 5.e

10. Match

1	Lygaeidae	a.	Cicada
2	Cicadidae	b.	Planthopper
3	Pyrrocorridae	c.	leaf hopper
4	Jassidae	d.	Red bug
5	Lophopidae	e.	true bug

Ans. 1. e, 2.a, 3.d, 4.c, 5.b

11. Match

1	Lamellate	a.	Heteroptera
2	Gula	b.	Scarabidae
3	Predator	c.	Whirling beetle
4	Gyrinidae	d.	Carabidae
5	Urogomphi	e.	Cicindelidae

Ans. 1. b, 2.a, 3.e, 4.c, 5.d

12. Match

1	Triangulin Larva	a.	Meloidae
2	fron and vertex	b.	Hydrophillidae
3	water scavenger beetle	c.	Scarabidae
4	dung beetle	d.	Snout
5	ground beetle	e.	Carabidae

Ans. 1. a, 2.d, 3.b, 4.c, 5.e

13. Match

1	Myrmeleontidae	a.	Raptorial legs
2	Neuropteran	b.	Conical pits
3	Raphidioptera	c.	Nerve wings
4	Siphonaptera	d.	Snakefly
5	Nepidae	e.	Laterally compressed body

Ans. 1.b, 2.c, 3.d, 4.e, 5.a

14. Match

1	Megaloptera	a.	Antilion
2	Neuroptera	b.	Green lacewing
3	Myrmeleontidae	c.	Owlfly
4	Ascalaphidae	d.	Carnivorous insect
5	Mantispidae	e.	Dabsonfly

Ans. 1. e, 2.b, 3.a, 4.c, 5.d

15. Match

1	Grylloblattodea	a.	Angel insects
2	Mantaphasmatodea	b.	Web spinners
3	Blattodea	c.	Rock Crawlers
4	Embioptera	d.	Cockroach
5	Zoraptera	e.	Gladiators

Ans. 1. c, 2.e, 3.d, 4.b, 5.a

16. Match

1	Cockroach	a.	Long and segmented
2	Grylloblattodea	b.	Asymmertical cerci
3	Mantaphasmatodea	c.	Short and unsegmented
4	Embioptera	d.	Short and many segmented
5	Zoraptera	e.	Cerci short, one-segmented

Ans. 1. d, 2.a, 3.e, 4.b, 5.c

17. Match

1	Megachilidae.	a.	*Camponotus compressus*
2	Xylocopidae	b.	Yellow jackets, Hornets
3	Apidae	c.	Leaf cutter bee
4	Vespidae	d.	Carpenter bee
5	Formicidae	e.	*Apismelifera*

Ans. 1. c, 2.d, 3.e, 4.b, 5.a

18. Match

1	Formic acid.	a.	*Bracon brevicornis*
2	Acid gland/ Dufour's gland	b.	*Trichogramma chilonis*
3	Pupal parasitoids	c.	Apidae
4	Egg parasitoids	d.	Formicidae
5	Larval parasitoids	e.	*Tretastichus sp*

Ans. 1. d, 2.c, 3.e, 4.b, 5.a

19. Match

1	Furculum	a.	Protura
2	Cone heads	b.	Collembola
3	Silver fish	c.	Diplura
4	Jumping bristle tails	d.	Thysanura
5	Pincer like cerci	e.	Archeognatha

Ans. 1. b, 2.a, 3.d, 4.e, 5.c

20. Match

1	Protura	a.	Japygids
2	Collembola	b.	Spring tails
3	Thysanura	c.	Telson tails
4	Archeognatha	d.	Jumping bristle tails
5	Diplura	e.	Silverfish

Ans. 1. c, 2.b, 3.e, 4.d, 5.a

21. Match

1.	Orthoptera	a.	Hamula
2.	Dermaptera	b.	Silk gland
3.	Embioptera	c.	Pterostigma
4.	Collembolan	d.	Cerci forceps like
5.	Odonata	e.	Tegmin
6.	Danaidae	f.	Milkweed Butterflies
7.	Hesperiidae	g.	Tiger Moths
8.	Nymphalidae	h.	Brush-footed Butterflies

9.	Papilionidae	i.	Silk Moths
10.	Pieridae	j.	Whites and Sulfurs Moths
11.	Arctiidae	k.	Skippers
12.	Noctuidae	l.	Underwing Moths
13.	Saturniidae	m.	Swallowtail Butterflies
14.	Sphingidae	n.	Hawk Moths

Ans. 1. e, 2.d, 3.b, 4.a, 5.c, 6.f, 7.k, 8.h, 9.m, 10.j, 11.g, 12.l, 13.i, 14.n

9.4 Fill up the blanks

1. The queen termite attaining enormous size after mating and obesity is known as ——————
2. In termites both —————— and —————— are sterile
3. —————— suture cause wing shedding in Isoptera
4. In Psocoptera lacinia is rod like and is known as ——————
5. Homeostasis is seen in ——————
6. Foreceps like cerci present in ——————
7. Wings similar in size, form and venetion in ——————
8. Other insects and arthropods living in termiterium are collectively called as ——————
9. Trial pheromone in termite is ——————
10. In Isoptera each group of individuals that perform the same function is called ——————
11. Hind wings are reduced and knobbed at the end, called ——————
12. Type of metamorphosis in diptera is ——————
13. In brachycera pupae is ——————
14. Ptilinium present in —————— part of insect body.
15. Larva of culicidae called ——————
16. Anopheles transmits ——————
17. Culex so transmit ——————
18. Anal paddle in mosquito helps in ——————
19. Pupa is active in ——————
20. —————— is blood sucking ectoparasite in cattle.
21. Bladder footed leg is found in ——————
22. Mesothorax is represented dorsally by —————— in bugs.

23. Ambient vein is present in ——————
24. Type of eggs laid by whitefly ——————
25. In Thysanoptera, ovipositor is saw like in ——————
26. Sound producing organ in cicadas ——————
27. Cotton stainer belongs to the family ——————
28. Loop veins are found in ——————
29. Cuneus is absent in ——————
30. Hemelytra with distinct corium, clavus and cuneus (a triangular apical piece of the basal part of forewing) is the characteristic of ——————
31. Wire worm lay eggs in ——————
32. Red flour beetle also called as ——————
33. Oily principle responsible for causes blisters is ——————
34. Triangulin larvae commonly observed in which family ——————
35. Family of cigarette beetle ——————
36. Family of sweet potato weevil ——————
37. Type of metamorphosis observed in blister beetle ——————
38. Family of lesser grain borer ——————
39. Damaging stage of white grubs ——————
40. Exposed part of abdomen of bruchids called as ——————
41. Stalked eggs (pedicellate type) are to avoid the —————————— and ——————
42. Laterally compressed insect is ——————
43. Other names of green lacewings ——————
44. Dobson and Alder flies are coming under which order ——————
45. Which adult fly resembles like a dragonfly ——————
46. Siphonaptera having which type of mouth parts ——————
47. Rhaphidioptera is derived from Greek words.the word *raphid* means ——————
48. Synonyms of Siphonaptera ——————
49. In owlflies —————— cell is present in wings
50. Other name of Caddisfly ——————
51. Head of blattodean insects is ——————

52. Ocelli are reduced to fenestrae in ————
53. The fore wings are sclerotized as ———— in Blattodea.
54. Cockroaches lay their eggs in ————
55. Each ootheca contains about ————
56. The members of family Corydidae are commonly called as ————
57. Compound eyes are reduced and the ocelli are absent in ————
58. Legs of ice crawlers are of ————
59. Ice crawlers generally feed on ————
60. Head in mantaphasmatids is of ————
61. Ovipositor is blunt and short in ————
62. Mantisphasmatids are also called as ————
63. The females of ———— produce egg pods by using sand and gland secretions
64. In ———— cerci asymmetrical in which left cerci is longer than right one which helps in holding the female during mating
65. Eggs and early instar nymphs in ———— are guarded by the female. (Sub social insect).
66. The adult antenna is moniliform in ————
67. The ovipositor is vestigial or absent in ————
68. ———— are also found in association with termites
69. ———— are generally fungivorous or necrophilous
70. Evaniidae wasp are parasitic on ———— of cockroach.
71. Sphecidae wasp made their nest with the help of ————
72. The first abdominal segment of Hymenopteran insect fused with metathorax is known as ————
73. Honey bee has ———— metamorphosis.
74. Pollination in onion is mainly done with the help of ————
75. Cenchri in metanotum is present in ————
76. White ants comes under the ———— order while red ants belongs to ———— order.
77. Honey bee is a social insect having ———— no of caste in their life cycle.
78. In Thysanura ———— abdominal segment elongated to form a median caudal filament.

79. Terminal 11th abdominal segment with a median telson is characteristic of ———

80. Snow fleas belong to the order ———
81. Only genus in the order protura that posses a tracheal system is ———
82. Silverfish posses ——— mandibular articulation.
83. Diplurans have ——— type of antennae.
84. Collembola and Protura comes under the class ———
85. Cone head insects belong to ———
86. In proturans malphigian tubules are represented as ———
87. Springing organ called furcula is present in ———
88. Rasping and sucking type mouthpart found in ———
89. Thick and leathery fore wing called ———
90. Ventral tube of collembolan located on ———
91. Anamorphous development is common in order ———
92. The anterior thicken portion of wing in dragon fly which is blood filled and surrounded by veins known as ———
93. Raptorial type of legs found in ———
94. Fringe wing is characteristic feature of order ———
95. Order: ——— ——— lacking in india.
96. Mustard saw fly comes under family ———
97. Fore tarsus modified with silk glands is found in ———
98. The front pairs of wings are thick and form a hard shell over the abdomen of ———
99. One or two rows of small spines are present on hind tibia is most important feature of family ———
100. Larvae can spin silk webs which are used to build the cases and to capture food from water ———
101. Obligate parasite insect with twisted wing ———
102. Wingless insect with body flattened laterally from side to side ———
103. Abdomens enlarge which makes it look like the stinger of a scorpion ———
104. Closely resemble to lace wings except for the presence of a pleated region of their hind wing, helping to fold over the abdomen ———

105. Cerci unjointed and modified into prominent horney forcep is character of order ————

106. Respiratory organ of order Odonata is in immature stage ————

107. Labial and maxillary palp absent in order ————

Answer key

1	2	3	4	5
Physogastry	Workers, soldiers	Humeral	pick	Isoptera
6	**7**	**8**	**9**	**10**
Dermaptera	Isoptera	Termitophiles	Caproic acid	Caste
11	**12**	**13**	**14**	**15**
Halter	Complete	Exarate	Head	wrigglers
16	**17**	**18**	**19**	**20**
malaria	Filariasis	movement	Pupa of mosquito (Tumbler)	Horn fly
21	**22**	**23**	**24**	**25**
Thrips	scutellum	hemiptera	elogated (Stalked)	Terebrantia
26	**27**	**28**	**29**	**30**
Tymbal	Pyrrhocoridae	Miridae	Lygaeidae	Miridae
31	**32**	**33**	**34**	**35**
soil	Meal worms	Cantharidin	Meloidae	Anobiidae
36	**37**	**38**	**39**	**40**
Apionidae (Brentidae)	Hyper metamorphosis	Bostrychidae	grub	Pygidium
41	**42**	**43**	**44**	**45**
Predation and cannibalism	Flea	Goldeneyes, Stinkflies, Aphidlions	Megaloptera	Ascalaphidae
46	**47**	**48**	**49**	**50**
Piercing and sucking type	Needle	Aphaniptera	Hypostigmal	Mothflies
51	**52**	**53**	**54**	**55**
Hypognathous	Blattodea	Tegmina	ootheca	16-24 eggs
56	**57**	**58**	**59**	**60**
Domino cockroach	Grylloblattodeans	cursorial type	mosses and other insects	hypognathous type
61	**62**	**63**	**64**	**65**
mantaphasmatids	Heelwalkers	mantophasmtids	embiids	embiids
66	**67**	**68**	**69**	**70**
Angel insects (Zorapera), termites (Isoptera)	angel insects	Zorapterans	Zorapterans	Ootheca
71	**72**	**73**	**74**	**75**
Muds	Propodeum	Complete	Honey Bee	Hymenoptera
76	**77**	**78**	**79**	**80**
Isoptera, Hymenoptera	3	11th	Protura	Collembolan

81	**82**	**83**	**84**	**85**
Eosentomon	Dicondylc	monilifom	Eliplua	Proturans
86	**87**	**88**	**89**	**90**
Papillae	Collembolan	Thrips	Tegmina	1st abdominal segment
91	**92**	**93**	**94**	**95**
Protura	Remigium	Preying mantid	Thysanoptera	Zoraptera, Grylloblattodea, Mantophas-matodea
96	**97**	**98**	**99**	**100**
Tenthredinidae	Embioptera	Elytra	Cicadellidae	Trichoptera
101	**102**	**103**	**104**	**105**
Strepsiptera	Siphonaptera	Mecoptera	Megaloptera	Dermaptera
106	**107**			
Rectal and Caudal gills	Hemiptera			

9.5. Quiz

Set -A

1. The term taxonomy was coined by
   a) De Candolle b) Theophrastus
   c) Pliny d) Linnaeus
2. New Systematics differs from Classical Systematics in employing
   a) Experimental Taxonomy
   b) Biochemical and Cytotaxonomy
   c) All biological parameters
   d) Numerical Taxonomy
3. Karyotaxonomy is a component of
   a) Cytotaxonomy b) Experimental Taxonomy
   c) Biochemical Taxonomy d) Numerical Taxonomy
4. The term taxon was given by
   a) Meyer b) Linnaeus
   c) Lamarck d) De Candolle
5. Phylogenetic system differs from a natural system in its stress on
   a) Anatomical details
   b) Physiological traits
   c) Morphological details
   d) Origin and evolutionary trend

6. A genus with a single species is
   a) Monotypic b) Typical
   c) Atypical d) Polytypic
7. Hierarchy of categories was introduced by
   a) Linnaeus b) De Candolle
   c) Bauhin d) John Ray
8. ICZN was adopted in
   a) 1960 b) 1970
   c) 1964 d) 1974
9. Lectotype is
   a) Duplicate of holotype
   b) Specimen described along with holotype
   c) Specimen cited by author without making one holotype
   d) Specimen selected from original material for nomenclature type when there is no holotype
10. Isotype is a specimen
   a) Duplicate of holotype
   b) Described alongwith holotype
   c) Nomenclature type when the original is missing
   d) Cited by author when there is no holotype
11. Neotype is
   a) Nomenclature type from original material
   b) Nomenclature type when the original material is missing
   c) One of the two or more specimens cited by author
   d) New species discovered by a scientist
12. Who developed the concept of phylogeny?
   a) Linnaeus b) Lamarck
   c) Hippocrates d) Ernst Haeckel
13. Classification reflecting the evolutionary inter relationships of organisms is called
   a) Phylogenetic classification b) Artificial classification
   c) Natural classification d) Numerical classification

14. Cytotaxonomy is a form of
   a) Classical systematics
   b) New systematics
   c) Morpho-systematics
   d) All the above
15. Classification based on chromosome study or organisms is
   a) Biochemical taxonomy
   b) Karyotaxonomy
   c) Numerical taxonomy
   d) Experimental taxonomy
16. The category of family is between
   a) Genus and species
   b) Order and genus
   c) Phylum and genus
   d) Kingdom and class
17. The highest category in taxonomy is
   a) Phylum
   b) Class
   c) Kingdom
   d) Species
18. The lowest category in taxonomic hierarchy is
   a) Phylum
   b) Subspecies
   c) Species
   d) Variety
19. Related species which are reproductively isolated but morphologically similar are called
   a) Allopatric
   b) Sympatric
   c) Sibling
   d) Morphospecies
20. Species having many subspecies are
   a) Monotypic/Microspecies
   b) Allopatric
   c) Sibling
   d) Polytypic/Macrospecies
21. Ectoganthus type of mothparts
   a) Firebract
   b) Snow flea
   c) Japygids
   d) Telsontails
22. A part rami present in which segment of collembola
   a) Hamula
   b) Tenaculum
   c) Retinaculum
   d) All the above
23. Statement 1 : Excretory gland absent in Diplura
   Statement 2 : Cephalic gland act as excretory glands in Diplura
   a) Both Statement 1 and statement 2 are True
   b) Both Statement 1 and statement 2 are False
   c) Statement 1 is True and statement 2 is False
   d) Statement 1 False and statement 2 is True

24. Statement 1 : Eyes are present in protura

    Statement 2 : Antenna absent in protura

    a) Both Statement 1 and statement 2 are True
    b) Both Statement 1 and statement 2 are False
    c) Statement 1 is True and statement 2 is False
    d) Statement 1 False and statement 2 is True

25. Statement 1 : Only apterygote order (Thysanura) having Median caudal filament

    Statement 2 : Only pterygote order (Ephemeroptera) having Median caudal filament

    a) Both Statement 1 and statement 2 are True
    b) Both Statement 1 and statement 2 are False
    c) Statement 1 is True and statement 2 is False
    d) Statement 1 False and statement 2 is True

26. Secondary copulatory organ present in Dragonfly

    a) 2nd sterna of adult Female
    b) 2nd sterna of adult male
    c) 3rd sterna of adult Female
    d) 3rd sterna of adult male

27. Statement 1 : External gills present in Anisoptera

    Statement 2 : Caudal gills present in Zygoptera

    a) Both Statement 1 and statement 2 are True
    b) Both Statement 1 and statement 2 are False
    c) Statement 1 is True and statement 2 is False
    d) Statement 1 False and statement 2 is True

28. Nymphs are aquatic

    a) Plecoptera b) Plectoptera
    c) Odonata d) All the above

29. Male Genetalia asymmetrical

    a) Web spinners b) Rock crawlers
    c) Grylloblatodea d) Both B & C

30. Pronymphs present in

    a) Phyllidae b) Ensifera
    c) Phasmidae d) All the Above

31. Tympanum located on either side of first abdomen segment
    a) Acrididae b) Gryllotalpidae
    c) Gryllidae d) Tettigonidae

32. Completely Herbivorous order
    a) Phasmida b) Lepidoptera
    c) Dermaptera d) Both A & B

33. Pick out wrong statement related to earwig Hindwing
    a) Tegmen b) Membranous
    c) Semicircular d) Ear like

34. Statement 1 : Cerci of earwig Foreceps like and Straight in Male
    Statement 2 : Cerci of earwig Foreceps like and Bowed in Female
    a) Both Statement 1 and statement 2 are True
    b) Both Statement 1 and statement 2 are False
    c) Statement 1 is True and statement 2 is False
    d) Statement 1 False and statement 2 is True

35. Smoky wings present in
    a) Embioptera b) Derampera
    c) Lepidoptera d) Hymenoptera

36. Fenestrate (Degenerated ocelli)
    a) Cockroach b) Preying Mantis
    c) Termites d) White ants

37. Frontal gland present in termite ————
    a) Queen b) King
    c) Worker d) Solider

38. Smallest order
    a) Embioptera b) Phasmida
    c) Zorapera d) Psocoptera

39. Wings are held roof like over the abdomen
    a) Psocoptera b) Mallophaga
    c) Siphunculata d) Hymenoptera

40. Body is dorsoventrally Flattened
    a) Psocoptera
    b) Mallophaga
    c) Siphunculata
    d) Both B & C
41. Among heteropteran only one family secrete Honey dew
    a) Reduviidae
    b) Cimicidae
    c) Miridae
    d) Tingidae
42. Pick out wrong statement
    a) Embolium absent – Miridae
    b) Embolium present – Anthocoridae
    c) Cuneus Absent – Lygaeidae
    d) Clavus absent – Coreidae
43. IN Exopterygota order Nympal stage is followed by prepupal and pupal stage which are analogous to the pupae of endopterygota
    a) Thysanura
    b) White fly
    c) Thrips
    d) Lacewing
44. Hypostigmal cell present in
    a) Ascalpidae
    b) Mantispidae
    c) Chrysopidae
    d) Mecoptera
45. Blood gills present in
    a) Chironomidae
    b) Asilidae
    c) Tabanidae
    d) Cecidomyiidae
46. Mouth Beard present in
    a) Chironomidae
    b) Asilidae
    c) Tabanidae
    d) Cecidomyiidae
47. Chest Bone or Sterna spatula present in
    a) Chironomidae
    b) Asilidae
    c) Tabanidae
    d) Cecidomyiidae
48. Ptilinium present in
    a) Gall midges
    b) Robber flies
    c) Mosquito
    d) House fly

49. Statement 1 : Trochanter absent in larval strepsipterans
Statement 2 : Trochanter absent in Adult strepsipterans
a) Both Statement 1 and statement 2 are True
b) Both Statement 1 and statement 2 are False
c) Statement 1 is True and statement 2 is False
d) Statement 1 False and statement 2 is True

50. Divided compound eye for arial vision and aquatic vision
a) Gyrinidae b) Dystiscidae
c) Cicindelidae d) Cassidae

Answer

1	**2**	**3**	**4**	**5**	**6**	**7**	**8**	**9**	**10**
a	c	a	a	d	a	a	c	d	a
11	**12**	**13**	**14**	**15**	**16**	**17**	**18**	**19**	**20**
b	d	a	b	b	b	c	c	c	d
21	**22**	**23**	**24**	**25**	**26**	**27**	**28**	**29**	**30**
a	a	a	d	a	b	d	d	d	b
31	**32**	**33**	**34**	**35**	**36**	**37**	**38**	**39**	**40**
a	a	a	b	a	a	d	c	a	d
41	**42**	**43**	**44**	**45**	**46**	**47**	**48**	**49**	**50**
d	d	c	a	a	b	d	d	d	a

SET-B

1. Fireflies belongs to
a) Diptera b) Coleoptera
c) Hymenoptera d) Lepidoptera

2. Photogenetic segment in adult Glow worms
a) 6th segment in Female
b) 5th segment in Male
c) 7th segment in Female
d) 6th and 7th segment in female

3. Body regions have metallic luster
a) Buprestidae b) Cassidae
c) Cerambycidae d) Elateridae

4. Larva are dorsal; spiny to which excreta and exuviae are attached forming a faecal shield
a) Buprestidae b) Cassidae
c) Cerambycidae d) Elateridae

5. Odd one out
   a) Sawflies b) Mayflies
   c) Robber fly d) Whitefly
6. Trochanter has two segmented
   a) Damselfly b) Dragonfly
   c) Ichneumonidaeflies d) All the above
7. Forewing has Two recurrent viens
   a) Ichneumonidae b) Braconidae
   c) Bethylidae d) Both A & B
8. Adult are not aggressive and no stings
   a) Megachilidae b) Sphecidae
   c) Xylocopidae d) Evaniidae
9. Odd one out
   a) Bombycidae b) Cochilididae
   c) Crambidae d) Papilionidae
10. Young larva resembles bird like excreta
    a) Papilionidae b) Lycaenidae
    c) Satyridae d) Arctiidae
11. Pick out wrong statement related to pronotum shape
    a) Saddle – Grasshopper b) Collar – Homoptera
    c) Sheild – Cockoarch d) Hood – Anobidae
12. Pick out Wrong statement
    a) Live in rotten wood – Zoraptera
    b) Live in silken nest – Web spinners
    c) Live in soil litter – Collembola
    d) None of above
13. Y shaped epicranial suture is not present
    a) Psocoptera b) Deramptera
    c) Mecoptera d) Zoraptera
14. Pick out wrong one related to sound production
    a) Acrididae – Femero alary Mechanism
    b) Sphingidae – Tympal
    c) Tettigonidae – Alary
    d) Gryllotalpidae – Humming sound

15. Pick out wrong statement related to Ovipositer shape
    a) Acrididae – Sword
    b) Gryllidae – Needle
    c) Tephritidae – Horny
    d) Ichneumonidae – Longer than body
16. Pick out wrong statement related to Manidible
    a) Caliper – Carabidae
    b) Sickle shape – Cicindeldidae
    c) Functional mandibles – Micropterygidae
    d) Membranous – Megaloptera
17. Statement 1 : Hindwing modified as has haltere in housefly and female of coccidae

    Statement 2 : Forewing modified as Haltere in Strepistera
    a) Both Statement 1 and statement 2 are True
    b) Both Statement 1 and statement 2 are False
    c) Statement 1 is True and statement 2 is False
    d) statement 1 False and statement 2 is True
18. Pick out wrong statement
    a) Nymph of coccidae – Crawlers
    b) Grub of Meloidae – Triangulin
    c) Larva of strepsiptera – Planidia
    d) Nymph of kerridae – Wriggler
19. Pick out wrong statement related to silk producing mechanism
    a) Psocoptera – Labial gland
    b) Embioptera – Basal tarsi of frontal legs
    c) Lepidoptera – Mandibular gland
    d) Neuroptera – Malphigian tubules
20. Pick out wrong statement related to Hindtibia
    a) Two pair of apical spur – Hesperidae
    b) Large mobile apical spur – Delphacidae
    c) Crown of large spines – Leaf footed bug
    d) Double row of spines – Ciccadellidae

21. Johnston organ absent in

a) Protura
b) Collembola
c) Diplura
d) All the above

22. Pick out wrong statement related to Crochets

a) Circular – Skipper (Hespiridae)
b) Semicircle – Noctuidae
c) Triangle – Pyralidae
d) Absent – Tenthredinidae(Saw Fly)

23. Pick out wrong statement

a) Female carry eggs in back – Hydrophilidae (Water Scavenger Beetle)
b) male carry eggs in back – Belostomatidae (Water Bug)
c) Faecial Shield – Tortise beetle
d) Female carry eggs in back – Belostomatidae (Water Bug)

24. Pick out wrong statement

a) Palmer organ – Mayfly
b) Pearman Organ – Book lice
c) Sempers organ – Mite
d) Muller organ – Acrididae

25. Pick out wrong statement

a) Hemiptera – Copeognatha
b) Mallophaga – Pthiraptera
c) Siphunculata – Anoplura
d) Thysanoptera – Physopoda

26. Statement 1 : Hind femur move forward & Backward in Embiodera
Statement 2 : Nymphs and adults running sidewise in delphacidae

a) Both Statement 1 and statement 2 are True
b) Both Statement 1 and statement 2 are False
c) Statement 1 is True and statement 2 is False
d) statement 1 False and statement 2 is True

27. Pronotum is large and it covers the head

a) Membracidae
b) Cicadellidae
c) Cercopidae
d) Lophopidae

28. The organ of respiration in collembola-

a) Tenaculum
b) Trachea
c) cuticle
d) Ventral tube

29. Eye patch is found in-
    a) Japygids b) White ants
    c) Snow flea d) Embiids
30. Autotomy is seen in-
    a) Leaf insect b) Stick insect
    c) Angel insect d) All of the above
31. In protura malpighian tubules are represented as-
    a) Pellicle b) Gills
    c) Papillae d) Telson
32. In collembola, gonopore is present on ———— abdominal segment and anus on ———— ———— abdominal segment –
    a) 5,6 b) 3,4
    c) 4,5 d) 4,6
33. Example of ecological indicators-
    a) Plecoptera b) Dictyoptera
    c) Orthoptera d) Coleoptera
34. The only pterygote order which has median terminal filament-
    a) Thysanura b) Phasmida
    c) Ephemeroptera d) Hymenotera
35. Imago and sub-imago stages are found in-
    a) Stone fly b) May fly
    c) Robber fly d) House fly
36. Sub-costa terminates into nodus in-
    a) Diptera b) Odonata,
    c) Hemiptera d) Lepidoptera
37. Example of rectal gills-
    a) Damsel fly b) Dragon fly
    c) Dobson fly d) Tumbler
38. In adults of Odonata, galea and lacinia are fused to form-
    a) Gula b) Mala
    c) Papillae d) Tube
39. Body of sub-imago is covered with fine set of hairs, called-
    a) Bristle b) Clothing hair
    c) Pellicle d) Plumose hair

40. Ribaga's organ found in-
   a) Bed bug b) Reduviid bug,
   c) Cow bug d) All
41. Comstock Kellog's gland is present in-
   a) Female acrididae b) Male acrididae
   c) Female arctiidae d) Male arctiidae
42. Monilliform antenna is found in-
   a) Trichoptera b) Mecoptera
   c) Zoraptera d) Siphonoptera
43. In many orthopterans, the newly hatched first instar nymphs are covered by loose cuticle are called-
   a) Sub-imago b) Protonymph
   c) Pronymph d) Cuticula
44. Smallest order-
   a) Zoraptera b) Plecoptera
   c) Grylloblattodea d) Ephemeroptera
45. Largest order-
   a) Coleoptera b) Lepidoptera
   c) Hymenoptera d) Diptera
46. Largest family in the order Coleoptera-
   a) Curculionidae b) Coccinellidae
   c) Bruchidae d) Scarabidae
47. Example of Mantophasmatodea-
   a) Phasmid b) Healwalker
   c) Rock crawler d) Mantispid fly
48. In Mantodea, eggs are filled in air filled solidified foam-
   a) Egg pod b) Egg sac
   c) Egg raft d) Spumaline
49. median caudal filament found in
   a) Thysanura b) Ephemeroptera
   c) Both d) None
50. Y-shaped epicranial suture found in
   a) Dermoptera b) Psocoptera
   c) Zoroptera d) All

Answer

1	2	3	4	5	6	7	8	9	10
b	d	a	b	c	d	a	d	d	a
11	**12**	**13**	**14**	**15**	**16**	**17**	**18**	**19**	**20**
d	d	c	b	a	d	d	d	c	c
21	**22**	**23**	**24**	**25**	**26**	**27**	**28**	**29**	**30**
d	c	d	c	a	c	a	d	c	b
31	**32**	**33**	**34**	**35**	**36**	**37**	**38**	**39**	**40**
c	a	a	c	b	b	b	b	c	a
41	**42**	**43**	**44**	**45**	**46**	**47**	**48**	**49**	**50**
a	c	c	a	a	a	b	d	c	d

SET - C

1. In cockroach, the degenerated ocelli are called-
   a) Fenestrae b) Frenate
   c) Stemmata d) All of the above
2. Link between orthopteroid and hemipteroid order-
   a) Zoraptera b) Thysanoptera
   c) Thysanura d) Diplura
3. The queen termite obtain enormous size after mating, and obesity is called-
   a) Physogastry b) Trophallaxis
   c) Fat body d) None of the above
4. In psocoptera, lacinia is called-
   a) Cootie b) Pick
   c) Nits d) File
5. Which exopterygote order that poses pre pupal and pupal stage like endopterygote-
   a) Lepidoptera b) Neuroptera
   c) Coleoptera d) Thysanoptera
6. In Hemiptera mesothorax is dorsally represented by-
   a) Scutellum b) Pseudoscutellum
   c) Scutum d) Epimeron
7. Respiratory siphon is present in-
   a) Tumbler b) Wriggler
   c) Water scorpion d) Mayfly naid

8. Repiratory horn is present in-
   a) Tumbler b) Wriggler
   c) Water scorpion d) Mayfly naid
9. In aphids, cornicles are present in which abdominal segment-
   a) 2-3 b) 3-4
   c) 4-5 d) 5-6
10. Vasiform orifice present in-
   a) Whitefly b) Aphids
   c) Jassids d) GLH
11. Link between Hemiptera and endopterygota-
   a) Thysanoptera b) Zoraptera
   c) Thysanura d) Diplura
12. Siphunculi or wax tube present in-
   a) Whitefly b) Aphids
   c) Jassids d) GLH
13. Conical pits are constructed by-
   a) Ground beetle b) Antlion grub
   c) Dung beetle d) Dragonfly
14. Vertical pits are constructed by-
   a) Ground beetle b) Antlion grub
   c) Dung beetle d) Dragonfly
15. The active first instar grub of blister beetle is known-
   a) Urogomphi b) Pygidium
   c) Triungulin d) Planidium
16. Insects walk in which fashion-
   a) Tripod locomotion b) Tetrapod locomotion
   c) Dipod locomotion d) Monopod locomotion
17. Order of wire worm-
   a) Cassididae b) Galerucidae
   c) Anobiidae d) Elateridae
18. Chest bone/ sterna spatula is found in-
   a) Adult ceciidomyidae, b) Larval cassididae,
   c) Adult cassididae, d) Larval ceciidomyidae

19. Blood gills are present in-
    a) Chironomid larvae b) Mayfly niad
    c) Damsel fly naid d) Dragonfly naid
20. Thyridium is present in-
    a) Siphonoptera b) Odonata
    c) Strepsiptera d) Trichoptera
21. Lepidopteran family that poses 4 functional legs-
    a) Nymphalidae b) Satyridae
    c) Pieridae d) Arctiidae
22. Clubbed antennae with a hook is present in-
    a) Rice skipper b) Sphinx moths
    c) Blue butterfly d) Blister beetle
23. Crochets are absent in-
    a) Eurema hecabe b) Papilio demoleus
    c) *Athalia lugens proxima* d) Ergolis marione
24. Pylifer is the sound producing organ of-
    a) Sphingidae b) Gryllidae
    c) Papilionidae d) Culicidae
25. Scansorial legs are found in-
    a) Mallophaga b) Psocoptera
    c) Siphunculata d) Psyllidae
26. Larvae of mosquito-
    a) Wriggler b) Tumbler
    c) Triungulin d) Ptillinum
27. Pupa of mosquito-
    a) Wriggler b) Tumbler
    c) Triungulin d) Ptillinum
28. Tuft of hairs forming mouth beard in-
    a) Horse fly b) Robber fly
    c) Hoverfly d) Tachinid fly
29. The larvae of Lepidoptera-
    a) Polypod b) Apod
    c) Eruciform d) Both a and c

30. Cannibalism is seen in-
    a) Earwig
    b) *Helicoverpa sp.*
    c) *Spodoptera frugiperda*
    d) All of the above
31. Pterostigmata is present in-
    a) Odonata
    b) Psocoptera
    c) Orthoptera
    d) Both a & b
32. Hellgrammatid is the larvae of –
    a) Dobson fly
    b) Dragon fly
    c) Damselfly
    d) Stone fly
33. Labial mask is seen in-
    a) Naids of dragonfly
    b) Naids of damselfly
    c) Naids of mayfly
    d) Both a & b
34. Tympanum is present on 1st abdominal segment in-
    a) Tettigonidae
    b) Acrididae
    c) Gryllotalpidae
    d) None
35. Tympanum is present on proximal end of foretibia in-
    a) Tettigonidae
    b) Acrididae
    c) Gryllotalpidae
    d) None
36. Biodiversity termed coined in 1980 by.....
    a) W.G. Rosen
    b) Wheeler
    c) Lahni
    d) Turner
37. Cyber taxonomy term was coined by
    a) Wheeler
    b) MS Mani
    c) Ernst Mayr
    d) Wilson
38. Which among the following is false?
    a) Cladistic classification only deals with monophyletic groups.
    b) Numerical taxonomy deals only with morphological characters
    c) The best phenetic classification is one constructed by considering single attribute
    d) Evolutionary classification is a synthesis of phonetic and phylogenetic principle.

39. When one taxon is replaced by another without branching during evolution, it is:
    a) Cladogenesis b) Anagenesis
    c) Morphogenesis d) None of the above
40. Which article in the ICZN deals with Law of Priority?
    a) Article 23 b) Article 33
    c) Article 53 d) Article13
41. The opposite sex specimen which is described along with the holotype
    a) Paratype b) Allotype
    c) Topotype d) Syntype
42. Which among the following is analogous structure:
    a) Wing of butterfly and Haltere in Dipteran
    b) Raptorial leg of preying mantis and Clasping leg in flea
    c) Wing of butterfly and wing of a bat
    d) Cursorial leg of cockroach and pectoral fin of whale.
43. The headquarters for the International Commission on Zoological Nomenclature is Located at :
    a) Singapore b) Moscow
    c) Geneva d) London
44. A species which is genetically and sexually different and are incapable of interbreeding:
    a) Sibling species b) Cryptic species
    c) Parapatric species d) Sympatric species
45. Which is false about PhyloCode?
    a) The first version of phyloCode was published in 2000.
    b) Theoretical foundation of the phyloCode was developed by de Queiroz and Gauthier
    c) It is a phylogenetic code of biological nomenclature, which is considered alternative to the Linnaean system
    d) It does not take common ancestry in consideration.
46. The objective of the ICZN Code was highlighted in the first edition in the year :
    a) 1961 b) 1964
    c) 1985 d) 1999

47. Consider the following statements:
    i) Linnaean system of binomial nomenclature is now becoming unsuitable to govern the naming of clade and species.
    ii) Phylocode is a newer system of biological nomenclature which considers the phylogenetics for nomenclature.
    a) Both statements are true
    b) Both statements are wrong
    c) First statement is wrong and second statement is true
    d) First statement is true and second is flase
48. Homology is
    a) Different origin and different function
    b) Same origin and same function
    c) Different origin and same function
    d) Similar origin but different function
49. Cohort exist between
    a) Sub class and super order
    b) Genus and species
    c) Family and genus
    d) None
50. A local population of a species of potentially inbreeding individuale at a given locality is called
    a) Race
    b) Deme
    c) Genus
    d) Order

Answer

1	**2**	**3**	**4**	**5**	**6**	**7**	**8**	**9**	**10**
a	a	a	b	d	a	b	a	d	a
11	**12**	**13**	**14**	**15**	**16**	**17**	**18**	**19**	**20**
a	b	b	a	c	a	d	d	a	d
21	**22**	**23**	**24**	**25**	**26**	**27**	**28**	**29**	**30**
a	a	c	a	c	a	b	b	d	d
31	**32**	**33**	**34**	**35**	**36**	**37**	**38**	**39**	**40**
d	a	d	b	a	a	a	c	b	a
41	**42**	**43**	**44**	**45**	**46**	**47**	**48**	**49**	**50**
b	c	d	b	d	a	a	d	a	b

9.6 Special organ/character

- Palmens organ ———— Ephemeroptera
- Mala ———— Odonata(Adult)
- Mask ———— Odonata, modififation of labium (Naids)
- Pronymph ———— Orthoptera

- Mullers organ ———— Acrididae
- Comstock kebleg gland ———— female acridid
- Y- shaped epicrianial suture ———— Dermaptera, pscoptera, Zoraptera
- Shaddle shape pronotum ———— Hodotermitidae, Acrididae
- Paddle shape wings ———— Zoraptera.
- Lacina (rod like pick) ———— Psocoptera
- Pearmans organ ———— Psocoptera
- Pulsatile organ ———— aquatic heteroptera
- Y-shaped vein (1A, 2A) ———— Fulgoroidae
- Embodium (sub equal segment) ———— Aleurodidae
- Breathing tube ———— Belostoatidae,Nepidae
- Respiratory horn ———— Nepidae egg
- Collar like pronotum ———— Pentotomodae
- Cyclomorphosis ———— Aphidae
- Repugnatorial gland ———— 2nd and 3rd abdominal segment
- Mushroom gland ———— cockroach
- Conglobate gland ———— cockroach
- Uricose gland ———— cockroach
- Frontal gland ———— termite
- Sterna gland ———— 3rd, 4th, 5th Mastotermas, 4th and 5th Hodotermus, 5th in other termites
- Fontonelle ———— Zoroptera
- Labial silk gland ———— Psocoptera
- Repugnatorial gland ———— hemiptera/heteroptera (Pentatomidae: adult-metathorax and nymph 6th to 7th abdominal segment
- Wax gland ———— homoptera
- Stink gland ———— 3rd abdominal segment in Cimicisae and inside metathorax Coreidae
- Cornical ———— 5th sement
- Dermal gland ———— lac insect
- Median caudal filament ———— Thysanura/Ephemeroptera
- Adult moulting ———— Thysanura/Ephemeroptera

- 10-12 annuli type antennae ———— odonata
- 4-segmented antennae ———— Colombola
- 9-segmented moniliform antennae ———— Zoraptera

Cerci

- Grylloblatodea ———— 8-segmented
- Orthoptera (grasshopper) ———— short/unsegmented
- Cricket ———— long/unsegmented
- Phasmida ———— short/unsegmented
- Cockroach ———— short/ many segmented

Assymetrical

- Male genetalia ———— Zoroptera /Grylloblatodea/phasmida
- Assymetrical cerci ———— embids

Malphigian tube

- Tettigonidae ———— Ampullae type
- Gryllidae ———— Uretes type

Mandible

- Tusk like epically pointed ———— orthoptera
- Tusk like ———— megaloptera
- Sickle shape ———— strepsiptera, cicindelidae,myrmeleontidae
- Dentate type ———— owl flies
- Caliper like ———— carabidae

Tentorium

- Imperforate ———— dermaptera
- Perforate ———— isoptera /cockroach

Some important Super Family

Hemiptera

Superfamily : Gerroidea (Family: Veliidae, Gerridae)

Superfamily : Nepoidea (Family: Belastomatidae, Nepidae

Superfamily : Reduviodea (Family: Reduviidae, Phymatidae, Enicocephalidae)

Superfamily : Meroidea (Family: Miridea Tingidae)
Superfamily : Cimicoidea (Family: Anthocoridae, Cimicidae-bed bug)
Superfamily : Pentatomoidea (Family: Pentatomidae, Scutelleridae)
Superfamily : Lygaeoidea (Family: Berytidae, Lygaeidae, Geocoridae)
Superfamily : Pyrrhocoroidea (Family: Pyrrhocoridae)
Superfamily : Coreoidea (Family: Coreidae-squash and leaf footed bug, Alydidae, Hyocephalidae, Rhopalidae, and Stenocephalidae)
Superfamily : Cicadoidea (Family: Cicadidae)
Superfamily : Cercopoidea (Cercopidea)
Superfamily : Membracoidea (Membracidae- tree hopper, Cicadellidae)
Superfamily : Fulgoroidea (Family: Delphacidae, Fulgoridae, Derbidae, Lophopidae)
Superfamily : Coccoidea (Family: Coccidae-sacle insect, Pseudococcidae-mealy bug, lacciferidae (Kerridae), Dactylopiidae-cochineal insect, Margarodidae)
Superfamily : Aphidoidea: (Family: Aphididae)
Superfamily : Phylloxeroidea (Phylloxeridae, Adelgidae)
Superfamily : Aleyrodoidea (Family: Aleyrodidae)

Hymenoptera

Superfamily : Tenthredinoidea (Family: Tenthredinidae -sawflies)
Superfamily : Ichneumonoidea (Family: Braconidae, Ichneumonidae)
Superfamily : Chalcidoidea (Family: Mymaridae-fairyflies, Trichogrammatidae, Eulophidae, Encyrtidae, Agaonidae-fig insects, Pteromalidae, Chalcididae
Superfamily : Vespoidea (Family: Vespidae)
Superfamily : Pompiloidea (Pompilidae)
Superfamily : Platygastroidea (Family: Scelionidae, Platygasteridae)
Superfamily : Evanioidea (Family: Aulacidae, Evaniidae-ensign wasps, Gasteruptiidae)
Superfamily : Chrysidoidea (Family: Bethylidae, Sclerogibbidae, Dryinidae)
Superfamily : Scolioidea (Family: Scoliidae-scoliid wasps)
Superfamily : Formicoidea (Family: Formicidae-ants)
Superfamily : Apoidea (Family: Colletidae - plasterer bees, yellow-faced bees, Andrenidae - mining bees, Halictidae-mining bees, Megachilidae-leaf-cutting bees, Apidae-bumblebees, honeybees, and digger, (Family: Sphecidae-sphecid wasps) or mining, bees, Anthophoridae- cuckoo bees, carpenter bees)

Lepidoptera

Superfamily : Noctuoidea (Family: Noctuidae-owlet moths, Arctiidae - tiger moths, Lymantriidae-tussock moths, Notodontidae-prominent moths)

Superfamily : Pyraloidea (Family: Pyralidae-snout, moths, Crambidae-webworms)

Superfamily : Geometroidea (Family: Geometridae-measuring worm, or inchworm, moths, Uraniidae-swallowtail moths)

Superfamily : Gelechioidea (Family: Gelechiidae-twirler moths)

Superfamily : Papilionoidea (Family: Lycaenidae-blues, coppers, hairstreaks, and metalmarks, Nymphalidae-brush-footed butterflies ,Pieridae-white, orange-tip, and sulfur butterflies, Papilionidae-swallowtails and parnassians)

Superfamily : Tineoidea (Family: Tineidae-clothes moths and other tineid moths, Psychidae-bagworms)

Superfamily : Gracillarioidea (Family: Gracillariidae and Douglasiidae)

Superfamily : Hesperioidea (Family: Hesperiidae-Skippers)

Superfamily : Bombycoidea (Family: Bombycidae-Silkworm Moths, Saturniidae -Giant Silkworm Moths, Sphingidae-Hawk, Or Sphinx, Moths)

Superfamily : Lasiocampoidea (Family: Lasiocampidae -Tent Caterpillar and Lappet Moths)

Superfamily : Yponomeutoidea (Family: Yponomeutidae-Ermine Moths)

Superfamily : Pterophoroidea (Family: Pterophoridae-Plume Moths)

Superfamily : Micropterigoidea (Family: Micropterigidae-Mandibulate Moths)

Coleoptera

Superfamily : Bostrichoidea (Family: Anobiidae-drugstore and death watch beetles, Bostrichidae-branch and twig borers, bostrichid beetles, horned powder post beetles, Dermestidae-skin beetles, dermestid beetles, Ptinidae-spider beetles)

Superfamily : Buprestoidea (Family: Buprestidae-metallic wood-boring beetles).

Superfamily : Chrysomeloidea (Family: Cerambycidae-long-horned beetles, Chrysomelidae-leaf beetles)

Superfamily : Coccinelloidea (Family: Coccinellidae-ladybird beetles, ladybugs, Corylophidae, Cryptophagidae-silken fungus beetles, Cucujidae-flat bark beetles)

Superfamily : Curculionoidea (Family: Attelabidae - leaf-rolling weevils, Belidae, Brentidae, Curculionidae-weevils)

Superfamily : Elateroidea (Family: Brachypsectridae, Cantharidae-soldier beetles, Cebrionidae, Cerophytidae, Drilidae, Elateridae-click beetles, Eucnemidae-false click beetles, Lampyridae-lightning bugs, fireflies, Lycidae-net-winged beetles, Phengodidae)

Superfamily : Histeroidea (Family: Histeridae-hister beetles and clown beetles)

Superfamily : Hydrophiloidea Sphaeritidae-false clown beetles, Synteliidae

Superfamily : Scarabaeoidea (Family: Scarabaeidae-scarb beetle, Lucanidae-stag beetle

Superfamily : Scirtoidea (Family : Clambidae-fringed-wing beetles, Decliniidae, Eucinetidae, Scirtidae, or)

Superfamily : Staphylinoidea (Family : Agyrtidae-primitive carrion beetles, Hydraenidae-minute moss beetles, Leiodidae-mammal-nest beetles, round fungus beetles, small carrion beetles, Ptiliidae-feather-winged beetles, Scydmaenidae- antlike stone beetles, Silphidae-large carrion beetles, burying beetles, Staphylinidae - rove beetles, Helodidae- Marsh beetle)

Superfamily : Tenebrionoidea (Family : Aderidae -antlike leaf beetles, Anthicidae -antlike flower beetles, Melandryidae-false darkling beetles, Meloidae-blister beetles, oil beetles, Oedemeridae-false blister beetles, Tenebrionidae-darkling beetles)

Diptera

Syperfamily : Culicoidea (Family: Culicidae, Dixidae – meniscus midges, Corethrellidae-frog-biting midges, Chaoboridae-phantom midges)

Superfamily : Chironomoidea (Chironomidae- Blood midges)

Syperfamily : Tephritoidea (Family: Tephritidae, Ulidiidae, Platystomatidae)

Syperfamily : Muscoidea (Family: Muscidae, Tachinidae, Anthomyiidae, Calliphoridae, Fanniidae-little house flies, Scathophagidae -dung flies)

Syperfamily : Carnoidea (Family: Acartophthalmidae, Australimyzidae· Braulidae-bee lice·Canacidae - beach flies·Carnidae· Chloropidae - fruit flies·Milichiidae)

Syperfamily : Opomyzoidea (Family: Agromyzidae-leaf miner flies, Anthomyzidae, Asteiidae, Aulacigastridae-sap flies, Neurochaetidae-upside-down flies, Opomyzidae, Teratomyzidae)

Syperfamily : Asiloidea (Family: Apioceridae-flower-loving flies, Apystomyiidae, Asilidae-robber flies, Bombyliidae-bee flies)

Superfamily : Syrphoidea (Families: Syrphidae-hover and drone flies, Pipunculidae - big-headed flies)

Superfamily : Ephydroidea (Families: Camillidae, Diastatidae - bog flies, Drosophilidae - vinegar and fruit flies, Ephydridae - shore flies, Mormotomyiidae)

Superfamily : Oestroidea (Families: Calliphoridae, Mesembrinellidae, Mystacinobiidae, Oestridae, Rhiniidae, Rhinophoridae, Sarcophagidae, Tachinidae)

Superfamily : Tipuloidea (family: Cylindrotomidae, Limoniidae, Pediciidae and Tipulidae)

Superfamily : Sciaroidea (Family: Cecidomyiidae – gall midges and wood midges, Diadocidiidae, Ditomyiidae, Keroplatidae, Lygistorrhinidae Sciaridae - dark-winged fungus gnats)

Superfamily : Tabanoidea (Family: Athericidae, Oreoleptidae, Pelecorhynchidae, Tabanidae -horse fly)

CLOSE RELATIVE OF CLASS INSECTA

Class : Crustacea

Lobsters, crabs, crayfish, shrimp, barnacles etc.

Body divided into two region, cephalothorax (head and thorax fused) and abdomen

Habitat: Terrestrial and Aquatic (mostly marine)

Antennae: Two pairs

Five pairs of leg

Respiration: Gills or body surface

Excretion: Green gland

Class : Arachnida

These are Chelicerates (with 6 pair of appendages; two pairs being mouthparts and four pairs legs) are characterized as having two distinct body regions, a cephlothorax and an abdomen

Habitat : Terrestrial

Antennae absent

Respiration by Trachea (ticks and mites) and book lungs in scorpions.
Excretion by Coxal gland or malpighian tubule

Class : Diplopoda

Millipedes

Habitat : Terrestrial

Millipedes are usually cylindrical (sometimes slightly flattened)

Two tagmata (head and trunk)

One pair of antennae

Two pairs of legs on most trunk segments (30 or more pairs total)

Mouthparts includes one pair of mandibles, and one pair of maxillae

Respiration by breathing tube

Excretion by malpighian tubul

Class: Chilopoda

Centipedes

Habitat : Terrestrial

Two tagmata (head and trunk)

One pair of antennae

One pair of legs per trunk segment (15 or more pairs total)

Mouthparts includes one pair of mandibles and two pairs of maxillae

Appendages on the first trunk segment are clawlike poison jaws or fangs for paralyze their prey

Respiration by breathing tube

Excretion by malpighian tubule